P9-AFQ-461

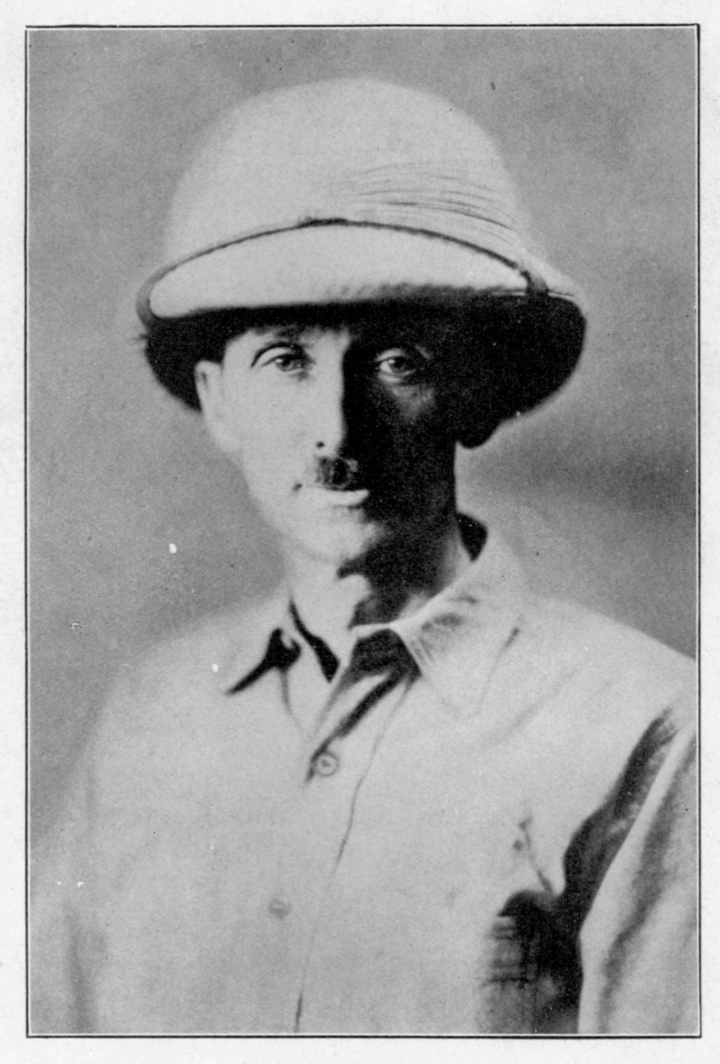

WILLIAM BEEBE

Author of Edge of the Jungle, Jungle Days, Gallapagos, World's End, The Arcturus Adventure, etc.

EDGE OF THE JUNGLE

BY

WILLIAM BEEBE

Honorary Curator of Birds and Director of the Tropical Research Station of the New York Zoological Society. Author of "Tropical Wild Life," "Monograph of the Pheasants," "Jungle Peace," etc.

GARDEN CITY NEW YORK

GARDEN CITY PUBLISHING COMPANY

1927

January, 1927

PRINTED IN THE UNITED STATES OF AMERICA

TO

THE BIRDS AND BUTTERFLIES,
THE ANTS AND TREE-FROGS
WHO HAVE TOLERATED ME IN
THEIR JUNGLE ANTE-CHAMBERS
I OFFER THIS VOLUME OF
FRIENDLY WORDS

NOTE

This second series of essays, following those in *Jungle Peace,* are republished by the kindness of the Editors of *The Atlantic Monthly, Harper's Magazine* and *House and Garden.*

With the exception of *A Tropic Garden* which refers to the Botanical Gardens of Georgetown, all deal with the jungle immediately about the Tropical Research Station of the New York Zoological Society, situated at Kartabo, at the junction of the Cuyuni and Mazaruni Rivers, in British Guiana.

For the accurate identification of the more important organisms mentioned, a brief appendix of scientific names has been prepared.

CONTENTS

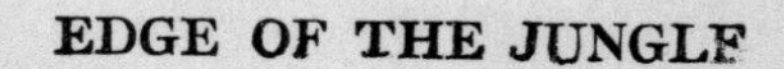

EDGE OF THE JUNGLE

"For the true scientific method is this:
To trust no statements without verification,
to test all things as rigorously as possible,
to keep no secrets, to attempt no monopolies,
to give out one's best modestly and plainly,
serving no other end but knowledge."

H. G. Wells.

I

THE LURE OF KARTABO

A HOUSE may be inherited, as when a wren rears its brood in turn within its own natal hollow; or one may build a new home such as is fashioned from year to year by gaunt and shadowy herons; or we may have it built to order, as do the drones of the wild jungle bees. In my case, I flitted like a hermit crab from one used shell to another. This little crustacean, living his oblique life in the shallows, changes doorways when his home becomes too small or hinders him in searching for the things which he covets in life. The difference between our estates was that the hermit crab sought only for food, I chiefly for strange new facts—which was a distinction as trivial as that he achieved his desires sideways and on eight legs, while I traversed my environment usually forward and generally on two.

The word of finance went forth and demanded the felling of the second growth around Kalacoon, and for the second time the land was given over to cutlass and fire. But again there was a halting in the affairs of man, and the rubber saplings were not planted or were smothered; and again the jungle smiled patiently through a knee-tangle of thorns and blossoms, and the charred clumps of razor-grass sent forth skeins of saws and hanks of living barbs.

I stood beneath the familiar cashew trees, which had yielded for me so bountifully of their crops of blossoms and hummingbirds, of fruit and of tanagers, and looked out toward the distant jungle, which trembled through the expanse of palpitating heat-waves; and I knew how a hermit crab feels when its home pinches, or is out of gear with the world. And, too, Nupee was dead, and the jungle to the south seemed to call less strongly. So I wandered through the old house for the last time, sniffing the agreeable odor of aged hypo still permeating the dark room, re-covering the empty stains of skins and traces of maps on the walls, and re-filling in my mind the vacant shelves. The vampires had returned to their chosen roost, the martins still swept through the corridors, and as I went down the hill, a moriche oriole sent a silver shaft of song after me from the sentinel palm, just as he had greeted me four years ago.

Then I gathered about me all the strange and unnamable possessions of a tropical laboratory—and moved. A wren reaches its home after hundreds of miles of fast aerial travel; a hermit crab achieves a new lease with a flip of his tail. Between these extremes, and in no less strange a fashion, I moved. A great barge pushed off from the Penal Settlement, piled high with my zoölogical Lares and Penates, and along each side squatted a line of paddlers,—white-garbed burglars and murderers, forgers and fighters,—while seated aloft on one of my ammunition trunks, with a microscope case and a camera close under his watchful eye, sat Case, King of the Warders, the biggest, blackest, and kindest-hearted man in the world.

Three miles up river swept my moving-van; and from the distance I could hear the half-whisper—which was yet a roar—of Case as he admonished his children. "Mon," he would say to a shirking, shrinking coolie second-story man, "mon, do you t'ink dis the time to sleep? What toughts have you in your bosom, dat you delay de Professor's household?" And then a chanty would rise, the voice of the leader quavering with that wild rhythm which had come down to him, a vocal heritage, through centuries of tom-toms and generations of savages striving for emotional expression. But the words were laughable or pathetic. I was adjured to

> "Blow de mon down with a bottle of rum,
> Oh, de mon—mon—blow de mon down."

Or the jungle reëchoed the edifying reiteration of

> "Sardines—and bread—OH!
> Sardines—and bread,
> Sardines—and bread—AND!
> Sardines—and bread."

The thrill that a whole-lunged chanty gives is difficult to describe. It arouses some deep emotional response, as surely as a military band, or the reverberating cadence of an organ, or a suddenly remembered theme of opera.

As my aquatic van drew up to the sandy landing-beach, I looked at the motley array of paddlers, and my mind went back hundreds of years to the first Spanish crew which landed here, and I wondered whether these pirates of early days had any fewer sins to their credit than Case's convicts—and I doubted it.

Across my doorstep a line of leaf-cutting ants was passing, each bearing aloft a huge bit of green leaf, or a long yellow petal, or a halberd of a stamen. A shadow fell over the line, and I looked up to see an anthropomorphic enlargement of the ants,—the convicts winding up the steep bank, each with cot, lamp, table, pitcher, trunk, or aquarium balanced on his head,—all my possessions suspended between earth and sky by the neck-muscles of worthy sinners. The first thing to be brought in was a great war-bag packed to bursting, and Number 214, with eight more years to serve, let it slide down his shoulder with a grunt—the self-same sound that I have heard from a Tibetan woman carrier, and a Mexican peon, and a Japanese porter, all of whom had in past years toted this very bag.

I led the way up the steps, and there in the doorway was a tenant, one who had already taken possession, and who now faced me and the trailing line of convicts with that dignity, poise, and perfect self-possession which only a toad, a giant grandmother of a toad, can exhibit. I, and all the law-breakers who followed, recognized the nine tenths involved in this instance and carefully stepped around. When the heavy things began to arrive, I approached diffidently, and half suggested, half directed her deliberate hops toward a safer corner. My feelings toward her were mingled, but altogether kindly,—as guest in her home, I could not but treat her with respect,—while my scientific soul revelled in the addition of *Bufo guttatus* to the fauna of this part of British Guiana. Whether flashing gold of oriole, or the blinking solemnity of a great toad, it mattered little—Kartabo had welcomed me with as propitious an omen as had Kalacoon.

Houses have distinct personalities, either bequeathed to them by their builders or tenants, absorbed from their materials, or emanating from the general environment. Neither the mind which had planned our Kartabo bungalow, nor the hands which fashioned it; neither the mahogany walls hewn from the adjoining jungle, nor the white-pine beams which had known many decades of snowy winters—none of these were obtrusive. The first had passed into oblivion, the second had been seasoned by sun and rain, papered by lichens, and gnawed and bored by tiny wood-folk into a neutral inconspicuousness as complete as an Indian's deserted *benab.* The wide verandah was open on all sides, and from the bamboos of the front compound one looked straight through the central hallway to bamboos at the back. It seemed like a happy accident of the natural surroundings, a jungle-bound cave, or the low rambling chambers of a mighty hollow tree.

No thought of who had been here last came to us that first evening. We unlimbered the creaky-legged cots, stiff and complaining after their three years' rest, and the air was filled with the clean odor of micaceous showers of naphthaline from long-packed pillows and sheets. From the rear came the clatter of plates, the scent of ripe papaws and bananas, mingled with the smell of the first fire in a new stove. Then I went out and sat on my own twelve-foot bank, looking down on the sandy beach and out and over to the most beautiful view in the Guianas. Down from the right swept slowly the Mazaruni, and from the left the Cuyuni, mingling with one wide expanse like a great rounded lake, bounded by solid jungle, with only Kalacoon and the Penal Settlement as tiny breaks in the wall of green.

The tide was falling, and as I sat watching the light grow dim, the water receded slowly, and strange little things floated past downstream. And I thought of the no less real human tide which long years ago had flowed to my very feet and then ebbed, leaving, as drift is left upon the sand, the convicts, a few scattered Indians, and myself. In the peace and quiet of this evening, time seemed a thing of no especial account. The great jungle trees might always have been lifeless emerald water-barriers, rather than things of a few centuries' growth; the rippleless water bore with equal disregard the last mora seed which floated past, as it had held aloft the keel of an unknown Spanish ship three centuries before. These men came up-river and landed on a little island a few hundred yards from Kartabo. Here they built a low stone wall, lost a few buttons, coins, and bullets, and vanished. Then came the Dutch in sturdy ships, cleared the islet of everything except the Spanish wall, and built them a jolly little fort intended to command all the rivers, naming it Kyk-over-al. To-day the name and a strong archway of flat Holland bricks survive.

In this wilderness, so wild and so quiet to-day, it was amazing to think of Dutch soldiers doing sentry duty and practising with their little bell-mouthed cannon on the islet, and of scores of negro and Indian slaves working in cassava fields all about where I sat. And this not fifty or a hundred or two hundred years ago, but about the year 1613, before John Smith had named New England, while the Hudson was still known as the Maurice, before the Mayflower landed with all our ancestors on board. For many years the story of this settlement and of the handful of neighboring sugar-plantations is one of privateer raids, capture, torture, slave-revolts, disease, bad government, and small profits, until we marvel at the perseverance of these sturdy Hollanders. From the records still extant, we glean here and there amusing details of the life which was so soon to falter and perish before the onpressing jungle. Exactly two hundred and fifty years ago one Hendrik Rol was appointed commander of Kyk-over-al. He was governor, captain, store-keeper and Indian trader, and his salary was thirty guilders, or about twelve dollars, a month—about what I paid my cook-boy.

The high tide of development at Kartabo came two hundred and three years ago, when, as we read in the old records, a Colony House was erected here. It went by the name of Huis Naby (the house near-by), from its situation near the fort. Kyk-over-al was now left to the garrison, while the commander and the civil servants lived in the new building. One of its rooms was used as a council chamber and church, while the lower floor was occupied by the company's store. The land in the neighborhood was laid out in building lots, with a view to establishing a town; it even went by the name of Stad Cartabo and had a tavern and two or three small houses, but never contained enough dwellings to entitle it to the name of town, or even village.

The ebb-tide soon began, and in 1739 Kartabo was deserted, and thirty years before the United States became a nation, the old fort on Kyk-over-al was demolished. The rivers and rolling jungle were attractive, but the soil was poor, while the noisome mud-swamps of the coast proved to be fertile and profitable.

Some fatality seemed to attach to all future attempts in this region. Gold was discovered, and diamonds, and to-day the wilderness here and there is powdering with rust and wreathing with creeping tendrils great piles of machinery. Pounds of gold have been taken out and hundreds of diamonds, but thus far the negro porkknocker, with his pack and washing-pan, is the only really successful miner.

The jungle sends forth healthy trees two hundred feet in height, thriving for centuries, but it reaches out and blights the attempts of man, whether sisal, rubber, cocoa, or coffee. So far the ebb-tide has left but two successful crops to those of us whose kismet has led us hither—crime and science. The concentration of negroes, coolies, Chinese and Portuguese on the coast furnishes an unfailing supply of convicts to the settlement, while the great world of life all about affords to the naturalist a bounty rich beyond all conception.

So here was I, a grateful legatee of past failures, shaded by magnificent clumps of bamboo, brought from Java and planted two or three hundred years ago by the Dutch, and sheltered by a bungalow which had played its part in the development and relinquishment of a great gold mine.

For a time we arranged and adjusted and shifted our equipment,—tables, books, vials, guns, nets, cameras and microscopes,—as a dog turns round and round before it composes itself to rest. And then one day I drew a long breath and looked about, and realized that I was at home. The newness began to pass from my little shelves and niches and blotters; in the darkness I could put my hand on flash or watch or gun; and in the morning I settled snugly into my woolen shirt, khakis, and sneakers, as if they were merely accessory skin.

In the beginning we were three white men and four servants—the latter all young, all individual, all picked up by instinct, except Sam, who was as inevitable as the tides. Our cook was too good-looking and too athletic to last. He had the reputation of being the fastest sprinter in Guiana, with a record, so we were solemnly told, of $9\frac{1}{5}$ seconds for the hundred—a veritable Mercury, as the last world's record of which I knew was $9\frac{3}{5}$. His stay with us was like the orbit of some comets, which make a single lap around the sun never to return, and his successor Edward, with unbelievably large and graceful hands and feet, was a better cook, with the softest voice and gentlest manner in the world.

But Bertie was our joy and delight. He too may be compared to a star—one which, originally bright, becomes temporarily dim, and finally attains to greater magnitude than before. Ultimately he became a fixed ornament of our culinary and taxidermic cosmic system, and whatever he did was accomplished with the most remarkable contortions of limbs and body. To watch him rake was to learn new anatomical possibilities; when he paddled, a surgeon would be moved to astonishment; when he caught butterflies, a teacher of physical culture would not have believed his eyes.

At night, when our servants had sealed themselves hermetically in their room in the neighboring thatched quarters, and the last squeak from our cots had passed out on its journey to the far distant goal of all nocturnal sounds, we began to realize that our new home held many more occupants than our three selves. Stealthy rustlings, indistinct scrapings, and low murmurs kept us interested for as long as ten minutes;

and in the morning we would remember and wonder who our fellow tenants could be. Some nights the bungalow seemed as full of life as the tiny French homes labeled, "*Hommes* 40: *Chevaux* 8," when the hastily estimated billeting possibilities were actually achieved, and one wondered whether it were not better to be the *cheval premier*, than the *homme quarantième*.

For years the bungalow had stood in sun and rain unoccupied, with a watchman and his wife, named Hope, who lived close by. The aptness of his name was that of the little Barbadian mule-tram which creeps through the coral-white streets, striving forever to divorce motion from progress and bearing the name Alert. Hope had done his duty and watched the bungalow. It was undoubtedly still there and nothing had been taken from it; but he had received no orders as to accretions, and so, to our infinite joy and entertainment, we found that in many ways it was not only near jungle, it *was* jungle. I have compared it with a natural cave. It was also like a fallen jungle-log, and we some of the small folk who shared its dark recesses with hosts of others. Through the air, on wings of skin or feathers or tissue membrane; crawling or leaping by night; burrowing underground; gnawing up through the great supporting posts; swarming up the bamboos and along the pliant curving stems to drop quietly on the shingled roof;—thus had the jungle-life come past Hope's unseeing eyes and found the bungalow worthy residence.

The bats were with us from first to last. We exterminated one colony which spent its inverted days clustered over the center of our supply chamber, but others came immediately and disputed the ownership of the dark room. Little chaps with great ears and nose-tissue of sensitive skin, spent the night beneath my shelves and chairs, and even my cot. They hunted at dusk and again at dawn, slept in my room and vanished in the day. Even for bats they were ferocious, and whenever I caught one in a butterfly-net, he went into paroxysms of rage, squealing in angry passion, striving to bite my hand and, failing that, chewing vainly on his own long fingers and arms. Their teeth were wonderfully intricate and seemed adapted for some very special diet, although beetles seemed to satisfy those which I caught. For once, the systematist had labeled them opportunely, and we never called them anything but *Furipterus horrens.*

In the evening, great bats as large as small herons swept down the long front gallery where we worked, gleaning as they went; but the vampires were long in coming, and for months we neither saw nor heard of one. Then they attacked our servants, and we took heart, and night after night exposed our toes, as conventionally accepted vampire-bait. When at last they found that the color of our skins was no criterion of dilution of blood, they came in crowds. For three nights they swept about us with hardly a whisper of wings, and accepted either toe or elbow or finger, or all three, and the cots and floor in the morning looked like an emergency hospital behind an active front. In spite of every attempt at keeping awake, we dropped off to sleep before the bats had begun, and did not waken until they left. We ascertained, however, that there was no truth in the belief that they hovered or kept fanning with their wings. Instead, they settled on the person with an appreciable flop and then crawled to the desired spot.

One night I made a special effort and, with bared arm, prepared for a long vigil. In a few minutes bats began to fan my face, the wings almost brushing, but never quite touching my skin. I could distinguish the difference between the smaller and the larger, the latter having a deeper swish, deeper and longer drawn-out. Their voices were so high and shrill that the singing of the jungle crickets seemed almost contralto in comparison. Finally, I began to feel myself the focus of one or more of these winged weasels. The swishes became more frequent, the returnings almost doubling on their track. Now and then a small body touched the sheet for an instant, and then, with a soft little tap, a vampire alighted on my chest. I was half sitting up, yet I could not see him, for I had found that the least hint of light ended any possibility of a visit. I breathed as quietly as I could, and made sure that both hands were clear. For a long time there was no movement, and the renewed swishes made me suspect that the bat had again taken flight. Not until I felt a tickling on my wrist did I know that my visitor had shifted and, unerringly, was making for the arm which I had exposed. Slowly it crept forward, but I hardly felt the pushing of the feet and pulling of the thumbs as it crawled along. If I had been asleep, I should not have awakened. It continued up my forearm and came to rest at my elbow. Here another long period of rest, and then several short, quick shifts of body. With my whole attention concentrated on my elbow, I began to imagine various sensations as my mind pictured the long, lancet tooth sinking deep into the skin, and the blood pumping up. I even began to feel the hot rush of my vital fluid over my arm, and then found that I had dozed for a moment and that all my sensations were imaginary. But soon a gentle tickling became apparent, and, in spite of putting this out of my mind and with increasing doubts as to the bat being still there, the tickling continued. It changed to a tingling, rather pleasant than otherwise, like the first stage of having one's hand asleep.

It really seemed as if this were the critical time. Somehow or other the vampire was at work with no pain or even inconvenience to me, and now was the moment to seize him, call for a lantern, and solve his supersurgical skill, the exact method of this vespertilial anæsthetist.

Slowly, very slowly, I lifted the other hand, always thinking of my elbow, so that I might keep all the muscles relaxed. Very slowly it approached, and with as swift a motion as I could achieve, I grasped at the vampire. I felt a touch of fur and I gripped a struggling, skinny wing; there came a single nip of teeth, and the wing-tip slipped through my fingers. I could detect no trace of blood by feeling, so turned over and went to sleep. In the morning I found a tiny scratch, with the skin barely broken; and, heartily disappointed, I realized that my tickling and tingling had been the preliminary symptoms of the operation.

Marvelous moths which slipped into the bungalow like shadows; pet tarantulas; golden-eyed gongasocka geckos; automatic, house-cleaning ants; opossums large and small; tiny lizards who had tongues in place of eyelids; wasps who had doorsteps and watched the passing from their windows;—all these were intimates of my laboratory table, whose riches must be spread elsewhere; but the sounds of the bungalow were common to the whole structure.

One of the first things I noticed, as I lay on my cot, was the new voice of the wind at night.

Now and then I caught a familiar sound,—faint, but not to be forgotten,—the clattering of palm fronds. But this came from Boomboom Point, fifty yards away (an outjutting of rocks where we had secured our first giant catfish of that name). The steady rhythm of sound which rose and fell with the breeze and sifted into my window with the moonbeams, was the gentlest *shusssssss*ing, a fine whispering, a veritable fern of a sound, high and crisp and wholly apart from the moaning around the eaves which arose at stronger gusts. It brought to mind the steep mountain-sides of Pahang, and windy nights which presaged great storms in high passes of Yunnan.

But these wonder times lived only through memory and were misted with intervening years, while it came upon me during early nights, again and again, that this was Now, and that into the hour-glass neck of Now was headed a maelstrom of untold riches of the Future—minutes and hours and sapphire days ahead—a Now which was wholly unconcerned with leagues and liquor, with strikes and salaries. So I turned over with the peace which passes all telling—the forecast of delving into the private affairs of birds and monkeys, of great butterflies and strange frogs and flowers. The seeping wind had led my mind on and on from memory and distant sorrows to thoughts of the joy of labor and life.

At half-past five a kiskadee shouted at the top of his lungs from the bamboos, but he probably had a nightmare, for he went to sleep and did not wake again for half-an-hour. The final swish of a bat's wing came to my ear, and the light of a fog-dimmed day slowly tempered the darkness among the dusty beams and rafters. From high overhead a sprawling tarantula tossed aside the shriveled remains of his night's banquet, the emerald cuirass and empty mahogany helmet of a long-horned beetle, which eddied downward and landed upon my sheet.

Immediately around the bungalow the bamboos held absolute sway, and while forming a very tangible link between the roof and the outliers of the jungle, yet no plant could obtain foothold beneath their shade. They withheld light, and the mat of myriads of slender leaves killed off every sprouting thing. This was of the utmost value to us, providing shade, clear passage to every breeze, and an absolute dearth of flies and mosquitoes. We found that the clumps needed clearing of old stems, and for two days we indulged in the strangest of weedings. The dead stems were as hard as stone outside, but the ax bit through easily, and they were so light that we could easily carry enormous ones, which made us feel like giants, though, when I thought of them in their true botanical relationship, I dwarfed in imagination as quickly as Alice, to a pigmy tottering under a blade of grass. It was like a Brobdingnagian game of jack-straws, as the cutting or prying loose of a single stem often brought several others crashing to earth in unexpected places, keeping us running and dodging to avoid their terrific impact. The fall of these great masts awakened a roaring swish ending in a hollow rattling, wholly unlike the crash and dull boom of a solid trunk. When we finished with each clump, it stood as a perfect giant bouquet, looking, at a distance, like a tuft of green feathery plumes, with the bungalow snuggled beneath as a toadstool is overshadowed by ferns.

Scores of the homes of small folk were uncovered by our weeding out—wasps, termites, ants, bees, wood-roaches, centipedes; and occasionally a small snake or great solemn toad came out from the débris at the roots, the latter blinking and swelling indignantly at this sudden interruption of his siesta. In a strong wind the stems bent and swayed, thrashing off every imperfect leaf and sweeping low across the roof, with strange scrapings and bamboo mutterings. But they hardly ever broke and fell. In the evening, however, and in the night, after a terrific storm, a sharp, unexpected *rat-tat-tat-tat,* exactly like a machine-gun, would smash in on the silence, and two or three of the great grasses, which perhaps sheltered Dutchmen generations ago, would snap and fall. But the Indians and Bovianders who lived nearby, knew this was no wind, nor yet weakness of stem, but Sinclair, who was abroad and who was cutting down the bamboos for his own secret reasons. He was evil, and it was well to be indoors with all windows closed; but further details were lacking, and we were driven to clothe this imperfect ghost with history and habits of our own devising.

The birds and other inhabitants of the bamboos, were those of the more open jungle,—flocks drifting through the clumps, monkeys occasionally swinging from one to another of the elastic tips, while toucans came and went. At evening, flocks of parrakeets and great black orioles came to roost, courting the safety which they had come to associate with the clearings of human pioneers in the jungle. A box on a bamboo stalk drew forth joyous hymns of praise from a pair of little God-birds, as the natives call the house-wrens, who straightway collected all the grass and feathers in the world, stuffed them into the tiny chamber, and after a time performed the ever-marvelous feat of producing three replicas of themselves from this trash-filled box. The father-parent was one concentrated mite of song, with just enough feathers for wings to enable him to pursue caterpillars and grasshoppers as raw material for the production of more song. He sang at the prospect of a home; then he sang to attract and win a mate; more song at the joy of finding wonderful grass and feathers; again melody to beguile his mate, patiently giving the hours and days of her body-warmth in instinct-compelled belief in the future. He sang while he took his turn at sitting; then he nearly choked to death trying to sing while stuffing a bug down a nestling's throat; finally, he sang at the end of a perfect nesting season; again, in hopes of persuading his mate to repeat it all, and this failing, sang in chorus in the wren quintette—I hoped, in

gratitude to us. At least from April to September he sang every day, and if my interpretation be anthropomorphic, why, so much the better for anthropomorphism. At any rate, before we left, all five wrens sat on a little shrub and imitated the morning stars, and our hearts went out to the little virile featherlings, who had lost none of their enthusiasm for life in this tropical jungle. Their one demand in this great wilderness was man's presence, being never found in the jungle except in an inhabited clearing, or, as I have found them, clinging hopefully to the vanishing ruins of a dead Indian's *benab,* waiting and singing in perfect faith, until the jungle had crept over it all and they were compelled to give up and set out in search of another home, within sound of human voices.

Bare as our leaf-carpeted bamboo-glade appeared, yet a select little company found life worth living there. The dry sand beneath the house was covered with the pits of ant-lions, and as we watched them month after month, they seemed to have more in common with the grains of quartz which composed their cosmos than with the organic world. By day or night no ant or other edible thing seemed ever to approach or be entrapped; and month after month there was no sign of change to imago. Yet each pit held a fat, enthusiastic inmate, ready at a touch to turn steam-shovel, battering-ram, bayonet, and gourmand. Among the first thousand-and-one mysteries of Kartabo I give a place to the source of nourishment of the sub-bungalow ant-lions.

Walking one day back of the house, I observed a number of small holes, with a little shining head just visible in each, which vanished at my approach. Looking closer, I was surprised to find a colony of tropical doodle-bugs. Straightway I chose a grass-stem and squatting, began fishing as I had fished many years ago in the southern states. Soon a nibble and then an angry pull, and I jerked out the irate little chap. He had the same naked bumpy body and the fierce head, and when two or three were put together, they fought blindly and with the ferocity of bulldogs.

To write of pets is as bad taste as to write in diary form, and, besides, I had made up my mind to have no pets on this expedition. They were a great deal of trouble and a source of distraction from work while they were alive; and one's heart was wrung and one's concentration disturbed at their death. But Kib came one day, brought by a tiny copper-bronze Indian. He looked at me, touched me tentatively with a mobile little paw, and my firm resolution melted away. A young coati-mundi cannot sit man-fashion like a bear-cub, nor is he as fuzzy as a kitten or as helpless as a puppy, but he has ways of winning to the human heart, past all obstacles.

The small Indian thought that three shillings would be a fair exchange; but I knew the par value of such stock, and Kib changed hands for three bits. A week later a thousand shillings would have seemed cheap to his new master. A coati-mundi is a tropical, arboreal raccoon of sorts, with a long, ever-wriggling snout, sharp teeth, eyes that twinkle with humor, and clawed paws which are more skilful than many a fingered hand. By the scientists of the world he is addressed as *Nasua nasua nasua*—which lays itself open to the twin ambiguity of stuttering Latin, or the echoes of a Princetonian football yell. The natural histories call him coati-mundi, while the Indian has by far the best of it, with the ringing, climactic syllables, *Kibihée!* And so, in the case of a being who has received much more than his share of vitality, it was altogether fitting to shorten this to Kib—Dunsany's giver of life upon the earth.

My heart's desire is to run on and tell many paragraphs of Kib; but that, as I have said, would be bad taste, which is one form of immorality. For in such things sentiment runs too closely parallel to sentimentality,—moderation becomes maudlinism,—and one enters the caste of those who tell anecdotes of children, and the latest symptoms of their physical ills. And the deeper one feels the joys of friendship with individual small folk of the jungle, the more difficult it is to convey them to others. And so it is not of the tropical mammal coati-mundi, nor even of the humorous Kib that I think, but of the soul of him galloping up and down his slanting log, of his little inner ego, which changed from a wild thing to one who would hurl himself from any height or distance into a lap, confident that we would save his neck, welcome him, and waste good time playing the game which he invented, of seeing whether we could touch his little cold snout before he hid it beneath his curved arms.

So, in spite of my resolves, our bamboo groves became the homes of numerous little souls of wild folk, whose individuality shone out and dominated the less important incidental casement, whether it happened to be feathers, or fur, or scales. It is interesting to observe how the Adam in one comes to the surface in the matter of names for pets. I know exactly the uncomfortable feeling which must have perturbed the heart of that pioneer of nomenclaturists, to be plumped down in the midst of "the greatest aggregation of animals ever assembled" before the time of Noah, and to be able to speak of them only as *this* or *that, he* or *she*. So we felt when inundated by a host of pets. It is easy to speak of the species by the lawful Latin or Greek name; we mention the specimen on our laboratory table by its common natural-history appellation. But the individual who touches our pity, or concern, or affection, demands a special title—usually absurdly inapt.

Soon, in the bamboo glade about our bungalow, ten little jungle friends came to live; and to us they will always be Kib and Gawain, George and Gregory, Robert and Grandmother, Raoul and Pansy, Jennie and Jellicoe.

Gawain was not a double personality—he was an intermittent reincarnation, vibrating between the inorganic and the essence of vitality. In a reasonable scheme of earthly things he filled the niche of a giant green tree-frog, and one of us seemed to remember that the Knight Gawain was enamored of green, and so we dubbed him. For the hours of daylight Gawain preferred the rôle of a hunched-up pebble of malachite; or if he could find a leaf, he drew eighteen purple vacuum toes beneath him, veiled his eyes with opalescent lids, and slipped from the mineral to the vegetable kingdom, flattened by masterly shading which filled the hollows and leveled the bumps; and the leaf became more of a leaf than it had been before Gawain was merged with it.

Night, or hunger, or the merciless tearing of sleep from his soul wrought magic and transformed him into a glowing, jeweled specter. He sprouted toes and long legs; he rose and inflated his sleek emerald frog-form; his sides blazed forth a mother-of-pearl waist-coat—a myriad mosaics of pink and blue and salmon and mauve; and from nowhere if not from the very depths of his throat, there slowly rose twin globes,—great eyes,—which stood above the flatness of his head, as mosques above an oriental city. Gone were the neutralizing lids, and in their place, strange upright pupils surrounded with vermilion lines and curves and dots, like characters of ancient illuminated Persian script. And with these appalling eyes Gawain looked at us, with these unreal, crimson-flecked globes staring absurdly from an expressionless emerald mask, he contemplated roaches and small grasshoppers, and correctly estimated their distance and activity. We never thought of demanding friendship, or a hint of his voice, or common froggish activities from Gawain. We were content to visit him now and then, to arouse him, and then leave him to disincarnate his vertebral outward phase into chlorophyll or lifeless stone. To muse upon his courtship or emotions was impossible. His life had a feeling of sphinx-like duration—Gawain as a tadpole was unthinkable. He seemed ageless, unreal, wonderfully beautiful, and wholly inexplicable.

II

A JUNGLE CLEARING

WITHIN six degrees of the Equator, shut in by jungle, on a cloudless day in mid-August, I found a comfortable seat on a slope of sandy soil sown with grass and weeds in the clearing back of Kartabo laboratory. I was shaded only by a few leaves of a low walnut-like sapling, yet there was not the slightest hint of oppressive heat. It might have been a warm August day in New England or Canada, except for the softness of the air.

In my little cleared glade there was no plant which would be wholly out of place on a New England country hillside. With debotanized vision I saw foliage of sumach, elm, hickory, peach, and alder, and the weeds all about were as familiar as those of any New Jersey meadow. The most abundant flowers were Mazaruni daisies, cheerful little pale primroses, and close to me, fairly overhanging the paper as I wrote, was the spindling button-weed, a wanderer from the States, with its clusters of tiny white blossoms bouqueted in the bracts of its leaves.

A few yards down the hillside was a clump of real friends—the rich green leaves of vervain, that humble little weed, sacred in turn to the Druids, the Romans, and the early Christians, and now brought inadvertently in some long-past time, in an overseas shipment, and holding its own in this breathing-space of the jungle. I was so interested by this discovery of a superficial northern flora, that I began to watch for other forms of temperate-appearing life, and for a long time my ear found nothing out of harmony with the plants. The low steady hum of abundant insects was so constant that it required conscious effort to disentangle it from silence. Every few seconds there arose the cadence of a passing bee or fly, the one low and deep, the other shrill and penetrating. And now, just as I had become wholly absorbed in this fascinating game,—the kind of game which may at any moment take a worth-while scientific turn,—it all dimmed and the entire picture shifted and changed. I doubt if any one who has been at a modern battle-front can long sit with closed eyes in a midsummer meadow and not have his blood leap as scene after scene is brought back to him. Three bees and a fly winging their way past, with the rise and fall of their varied hums, were sufficient to renew vividly for me the blackness of night over the sticky mud of Souville, and to cloud for a moment the scent of clover and dying grass, with that terrible sickly sweet odor of human flesh in an old shell-hole. In such unexpected ways do we link peace and war—suspending the greatest weights of memory, imagination, and visualization on the slenderest cobwebs of sound, odor, and color.

But again my bees became but bees—great, jolly, busy yellow-and-black fellows, who blundered about and squeezed into blossoms many sizes too small for them. Cicadas tuned up, clearing their drum-heads, tightening their keys, and at last rousing into the full swing of their ecstatic theme. And my relaxed, uncritical mind at present recorded no difference between the sound and that which was vibrated from northern maples. The tamest bird about me was a big yellow-breasted white-throated flycatcher, and I had seen this Melancholy Tyrant, as his technical name describes him, in such distant lands that he fitted into the picture without effort.

White butterflies flitted past, then a yellow one, and finally a real Monarch. In my boy-land, smudgy specimens of this were pinned, earnestly but asymetrically, in cigar-boxes, under the title of *Danais archippus.* At present no reputable entomologist would think of calling it other than *Anosia plexippus,* nor should I; but the particular thrill which it gave to-day was that this self-same species should wander along at this moment to mosaic into my boreal muse.

After a little time, with only the hum of the bees and the staccato cicadas, a double deceit was perpetrated, one which my sentiment of the moment seized upon and rejoiced in, but at which my mind had to conceal a smile and turn its consciousness quickly elsewhere, to prevent an obtrusive reality from dimming this last addition to the picture. The gentle, unmistakable, velvet warble of a bluebird came over the hillside, again and again; and so completely absorbed and lulled was I by the gradual obsession of being in the midst of a northern scene, that the sound caused not the slightest excitement, even internally and mentally. But the sympathetic spirit who was directing this geographic burlesque overplayed, and followed the soft curve of audible wistfulness with an actual bluebird which looped across the open space in front. The spell was broken for a moment, and my subconscious autocrat thrust into realization the instantaneous report—apparent blue-bird call is the note of a small flycatcher and the momentary vision was not even a mountain bluebird but a red-breasted blue chatterer! So I shut my eyes very quickly and listened to the soft calls, which alone would have deceived the closest analyzer of bird songs. And so for a little while longer I still held my picture intact, a magic scape, a hundred yards square and an hour long, set in the heart of the Guiana jungle.

And when at last I had to desert Canada, and relinquish New Jersey, I slipped only a few hundred miles southward. For another twenty minutes I clung to Virginia, for the enforced shift was due to a great Papilio butterfly which stopped nearby and which I captured with a lucky sweep of my net. My first thought was of the Orange-tree Swallow-tail, *née Papilio cresphontes*. Then the first lizards appeared, and by no stretch of my willing imagination could I pretend that they were newts, or fit the little emerald scales into a New England pasture. And so I chose for a time to live again among the Virginian butterflies and mockingbirds, the wild roses and the jasmine, and the other splendors of memory which a single butterfly had unloosed.

As I looked about me, I saw the flowers and detected their fragrance; I heard the hum of bees and the contented chirp of well-fed birds; I marveled at great butterflies flapping so slowly that it seemed as if they must have cheated gravitation in some subtle way to win such lightness and disregard of earth-pull. I heard no ugly murmur of long hours and low wages; the closest scrutiny revealed no strikes or internal clamorings about wrongs; and I unconsciously relaxed and breathed more deeply at the thought of this nature world, moving so smoothly, with directness and simplicity as apparently achieved ideals.

Then I ceased this superficial glance and looked deeper, and without moralizing or dragging in far-fetched similes or warnings, tried to comprehend one fundamental reality in wild nature—the universal acceptance of opportunity. From this angle it is quite unimportant whether one believes in vitalism (which is vitiating to our "will to prove"), or in mechanism (whose name itself is a symbol of ignorance, or deficient vocabulary, or both). Evolution has left no chink or crevice unfilled, unoccupied, no probability untried, no possibility unachieved.

The nearest weed suggested this trend of thought and provided all I could desire of examples; but the thrill of discovery and the artistic delight threatened to disturb for the time my solemn application of these ponderous truisms. The weed alongside had had a prosperous life, and its leaves were fortunate in the unadulterated sun and rain to which they had access. At the summit all was focusing for the consummation of existence: the little blossoms would soon open and have their one chance. To all the winds of heaven they would fling out wave upon wave of delicate odor, besides enlisting a subtle form of vibration and refusing to absorb the pink light —thereby enhancing the prospects of insect visitors, on whose coming the very existence of this race of weeds depended.

Every leaf showed signs of attack: scallops cut out, holes bored, stains of fungi, wreaths of moss, and the insidious mazes of leaf-miners. But, like an old-fashioned ship of the line which wins to port with the remnants of shot-ridden sails, the plant had paid toll bravely, although unable to defend itself or protect its tissues; and if I did not now destroy it, which I should assuredly not do, this weed would justify its place as a worthy link in the chain of numberless generations, past and to come.

More complex, clever, subtle methods of attack transcended those of the mere devourer of leaf-tissue, as radically as an inventor of most intricate instruments differs from the plodding tiller of the soil. In the center of one leaf, less disfigured than some of its fellows, I perceived four tiny ivory spheres, a dozen of which might rest comfortably within the length of an inch. To my eye they looked quite smooth, although a steady oblique gaze revealed hints of concentric lines. Before the times of Leeuwenhoek I should perhaps have been unable to see more than this, although, as a matter of fact, in those happy-go-lucky days my ancestors would doubtless have trounced me soundly for wasting my time on such useless and ungodly things as butterfly eggs. I thought of the coming night when I should sit and strain with all my might, striving, without the use of my powerful stereos, to separate from translucent mist of gases the denser nucleus of the mighty cosmos in Andromeda. And I alternately bemoaned my human limitation of vision, and rejoiced that I could focus clearly, both upon my butterfly eggs a foot away, and upon the spiral nebula swinging through the ether perhaps four hundred and fifty light-years from the earth.

I unswung my pocket-lens,—the infant of the microscope,—and my whole being followed my eyes; the trees and sky were eclipsed, and I hovered in mid-air over four glistening Mars-like planets—seamed with radiating canals, half in shadow from the slanting sunlight, and silhouetted against pure emerald. The sculpturing was exquisite. Near the north poles which pointed obliquely in my direction, the lines broke up into beads, and the edges of these were frilled and scalloped; and here again my vision failed and demanded still stronger binoculars. Here was indeed complexity: a butterfly, one of those black beauties, splashed with jasper and beryl, hovering nearby, with taste only for liquid nectar, yet choosing a little weed devoid of flower or fruit on which to deposit her quota of eggs. She neither turned to look at their beauties nor trusted another batch to this plant. Somehow, someway, her caterpillar wormhood had carried, through the mummified chrysalid and the reincarnation of her present form, knowledge of an earlier, infinitely coarser diet.

Together with the pure artistic joy which was stirred at the sight of these tiny ornate globes, there was aroused a realization of complexity, of helpless, ignorant achievement; the butterfly blindly pausing in her flower-to-flower fluttering —a pause as momentous to her race as that of the slow daily and monthly progress of the weed's struggle to fruition.

I took a final glance at the eggs before returning to my own larger world, and I detected a new complication, one which left me with feelings too involved for calm scientific contemplation. As if a Martian should suddenly become visible to an astronomer, I found that one of the egg planets was inhabited. Perched upon the summit—quite near the north pole—was an insect, a wasp, much smaller than the egg itself. And as I looked, I saw it at the climax of its diminutive life; for it reared up, resting on the tips of two legs and the iridescent wings, and sunk its ovipositor deep into the crystalline surface. As I watched, an egg was deposited, about the latitude of New York, and with a tremor the tiny wasp withdrew its instrument and rested.

On the same leaf were casually blown specks of dust, larger than the quartette of eggs. To the plant the cluster weighed nothing, meant nothing more than the dust. Yet a moment before they contained the latent power of great harm to the future growth of the weed—four lusty caterpillars would work from leaf to leaf with a rapidity and destructiveness which might, even at the last, have sapped the maturing seeds. Now, on a smaller scale, but still within the realm of insect life, all was changed—the plant was safe once more and no caterpillars would emerge. For the wasp went from sphere to sphere and inoculated every one with the promise of its kind. The plant bent slightly in a breath of wind, and knew nothing; the butterfly was far away to my left, deep-drinking in a cluster of yellow cassia; the wasp had already forgotten its achievement, and I alone—an outsider, an interloper—observed, correlated, realized, appreciated, and—at the last—remained as completely ignorant as the actors themselves of the real driving force, of the certain beginning, of the inevitable end. Only a momentary cross-section was vouchsafed, and a wonder and a desire to know fanned a little hotter.

I had far from finished with my weed: for besides the cuts and tears and disfigurements of the leaves, I saw a score or more of curious berry-like or acorn-like growths, springing from both leaf and stem. I knew, of course, that they were insect-galls, but never before had they meant quite so much, or fitted in so well as a significant phenomenon in the nexus of entangling relationships between the weed and its environment. This visitor, also a minute wasp of sorts, neither bit nor cut the leaves, but quietly slipped a tiny egg here and there into the leaf-tissue.

And this was only the beginning of complexity. For with the quickening of the larva came a reaction on the part of the plant, which, in defense, set up a greatly accelerated growth about the young insect. This might have taken the form of some distorted or deformed plant organ—a cluster of leaves, a fruit or berry or tuft of hairs, wholly unlike the characters of the plant itself. My weed was studded with what might well have been normal seed-fruits, were they not proved nightmares of berries, awful pseudo-fruits sprouting from horridly impossible places. And this excess of energy, expressed in tumorous outgrowths, was all vitally useful to the grub—

just as the skilful jiu-jitsu wrestler accomplishes his purpose with the aid of his opponent's strength. The insect and plant were, however, far more intricately related than any two human competitors: for the grub in turn required the continued health and strength of the plant for its existence; and when I plucked a leaf, I knew I had doomed all the hidden insects living within its substance.

The galls at my hand simulated little acorns, dull greenish in color, matching the leaf-surface on which they rested, and rising in a sharp point. I cut one through and, when wearied and fretted with the responsibilities of independent existence, I know I shall often recall and envy my grub in his palatial parasitic home. Outside came a rather hard, brown protective sheath; then the main body of the gall, of firm and dense tissue; and finally, at the heart, like the Queen's chamber in Cheops, the irregular little dwelling-place of the grub. This was not empty and barren; but the blackness and silence of this vegetable chamber, this architecture fashioned by the strangest of builders for the most remarkable of tenants, was filled with a nap of long, crystalline hairs or threads like the spun-glass candy in our

Christmas sweetshops—white at the base and shading from pale salmon to the deepest of pinks. This exquisite tapestry, whose beauties were normally forever hidden as well from the blind grub as from the outside world, was the ambrosia all unwittingly provided by the antagonism of the plant; the nutrition of resentment, the food of defiance; and day by day the grub gradually ate his way from one end to the other of his suite, laying a normal, healthful physical foundation for his future aerial activities.

The natural history of galls is full of romance and strange unrealities, but to-day it meant to me only a renewed instance of an opportunity seized and made the most of; the success of the indirect, the unreasonable—the long chance which so few of us humans are willing to take, although the reward is a perpetual enthusiasm for the happening of the moment, and the honest gambler's joy for the future. How much more desirable to acquire merit as a footless grub in the heart of a home, erected and precariously nourished by a worthy opponent, with a future of unnumbered possibilities, than to be a queen-mother in nest or hive—cared-for, fed, and cleansed by a host of slaves, but with less prospect of change or of adventure than an average toadstool.

Thus I sat for a long time, lulled by similitudes of northern plants and bees and birds, and then gently shifted southward a few hundred miles, the transition being smooth and unabrupt. With equal gentleness the dead calm stirred slightly and exhaled the merest ghost of a breeze; it seemed as if the air was hardly in motion, but only restless: the wings of the bees and the fly-catcher might well have caused it. But, judged by the sequence of events, it was the almost imperceptible signal given by some great Jungle Spirit, who had tired of playing with my dreams and pleasant fancies of northern life, and now called upon her legions to disillusion me. And the response was immediate. Three great shells burst at my very feet,—one of sound, one of color, and the third of both plus numbers,—and from that time on, tropical life was dominant whichever way I looked. That is the way with the wilderness, and especially the tropical wilderness—to surprise one in the very field with which one is most familiar. While in my own estimation my chief profession is ignorance, yet I sign my passport applications and my jury evasions as Ornithologist. And now this playful Spirit of the Jungle permitted me to meditate cheerfully on my ability to compare the faunas of New York and Guiana, and then proceeded to startle me with three salvos of birds, first physically and then emotionally.

From the monotone of under-world sounds a strange little rasping detached itself, a reiterated, subdued scraping or picking. It carried my mind instantly to the throbbing theme of the Niebelungs, onomatopoetic of the little hammers forever busy in their underground work. I circled a small bush at my side, and found that the sound came from one of the branches near the top; so with my glasses I began a systematic search. It was at this propitious moment, when I was relaxed in every muscle, steeped in the quiet of this hillside, and keen on discovering the beetle, that the first shell arrived. If I had been less absorbed I might have heard some distant chattering or calling, but this time it was as if a Spad had shut off its power, volplaned, kept ahead of its own sound waves, and bombed me. All that actually happened was that a band of little parrakeets flew down and alighted nearby.

When I discovered this, it seemed a disconcerting anti-climax, just as one can make the bravest man who has been under rifle-fire flinch by spinning a match swiftly past his ear.

I have heard this sound of parrakeet's wings, when the birds were alighting nearby, half a dozen times; but after half a hundred I shall duck just as spontaneously, and for a few seconds stand just as immobile with astonishment. From a volcano I expect deep and sinister sounds; when I watch great breakers I would marvel only if the accompanying roar were absent; but on a calm sunny August day I do not expect a noise which, for suddenness and startling character, can be compared only with a tremendous flash of lightning. Imagine a wonderful tapestry of strong ancient stuff, which had only been woven, never torn, and think of this suddenly ripped from top to bottom by some sinister, irresistible force.

In the instant that the sound began, it ceased; there was no echo, no bell-like sustained overtones; both ends were buried in silence. As it came to-day it was a high tearing crash which shattered silence as a Very light destroys darkness; and at its cessation I looked up and saw twenty little green figures gazing intently down at me, from so small a sapling that their addition almost doubled the foliage. That their small wings could wring such a sound from the fabric of the air was unbelievable. At my first movement, the flock leaped forth, and if their wings made even a rustle, it was wholly drowned in the chorus of chattering cries which poured forth unceasingly as the little band swept up and around the sky circle. As an alighting morpho butterfly dazzles the eyes with a final flash of his blazing azure before vanishing behind the leaves and fungi of his lower surface, so parrakeets change from screaming motes in the heavens to silence, and then to a hurtling, roaring boomerang, whose amazing unexpectedness would distract the most dangerous eyes from the little motionless leaf-figures in a neighboring tree-top.

When I sat down again, the whole feeling of the hillside was changed. I was aware that my weed was a northern weed only in appearance, and I should not have been surprised to see my bees change to flies or my lizards to snakes—tropical beings have a way of doing such things.

The next phenomenon was color,—unreal, living pigment,—which seemed to appeal to more than one sense, and which satisfied, as a cooling drink or a rare, delicious fragrance satisfies. A medium-sized, stocky bird flew with steady wing-beats over the jungle, in black silhouette against the sky, and swung up to an outstanding giant tree which partly overhung the edge of my clearing. The instant it passed the zone of green, it flashed out brilliant turquoise, and in the same instant I recognized it and reached for my gun. Before I retrieved the bird, a second, dull and dark-feathered, flew from the tree. I had watched it for some time, but now, as it passed over, I saw no yellow and knew it too was of real scientific interest to me; and with the second barrel I secured it. Picking up my first bird, I found that it was not turquoise, but beryl; and a few minutes later I was certain that it was aquamarine; on my way home another glance showed the color of forget-me-nots on its plumage, and as I looked at it on my table, it was Nile green. Yet the feathers were painted in flat color, without especial sheen or iridescence, and when I finally analyzed it, I found it to be a delicate calamine blue. It actually had the appearance of a too strong color, as when a glistening surface reflects the sun. From beak to tail it threw off this glowing hue, except for its chin and throat, which were a limpid amaranth purple; and the effect on the excited rods and cones in one's eyes was like the power of great music or some majestic passage in the Bible. You, who think my similes are overdone, search out in the nearest museum the dustiest of purple-throated cotingas, —*Cotinga cayana*,—and then, instead, berate me for inadequacy.

Sheer color alone is powerful enough, but when heightened by contrast, it becomes still more effective, and I seemed to have secured, with two barrels, a cotinga and its shadow. The latter was also a full-grown male cotinga, known to a few people in this world as the dark-breasted mourner (*Lipaugus simplex*). In general shape and form it was not unlike its cousin, but in color it was its shadow, its silhouette. Not a feather upon head or body, wings or tail showed a hint of warmth, only a dull uniform gray; an ash of a bird, living in the same warm sunlight, wet by the same rain, feeding on much the same food, and claiming relationship with a blazing-feathered turquoise. There is some very exact and very absorbing reason for all this, and for it I search with fervor, but with little success. But we may be certain that the causes of this and of the host of other unreasonable realities which fill the path of the evolutionist with never-quenched enthusiasm, will extend far beyond the colors of two tropical birds. They will have something to do with flowers and with bright butterflies, and we shall know why our "favorite color" is more than a whim, and why the Greeks may not have been able to distinguish the full gamut of our spectrum, and why rainbows are so narrow to our eyes in comparison to what they might be.

Finally, there was thrown aside all finesse, all delicacy of presentation, and the last lingering feeling of temperate life and nature was erased. From now on there was no confusion of zones, no concessions, no mental palimpsest of resolving images. The spatial, the temporal,—the hill-side, the passing seconds,—the vibrations and material atoms stimulating my five senses, all were tropical, quickened with the unbelievable vitality of equatorial life. A rustling came to my ears, although the breeze was still little more than a sensation of coolness. Then a deep whirr sounded overhead, and another, and another, and with a rush a dozen great toucans were all about me. Monstrous beaks, parodies in pastels of unheard-of blues and greens, breasts which glowed like mirrored suns,—orange overlaid upon blinding yellow,—and at every flick of the tail a trenchant flash of intense scarlet. All these colors set in frames of jet-black plumage, and suddenly hurled through blue sky and green foliage, made the hillside a brilliant moving kaleidoscope.

Some flew straight over, with several quick flaps, then a smooth glide, flaps and glide. A few banked sharply at sight of me, and wheeled to right or left. Others alighted and craned their necks in suspicion; but all sooner or later disappeared eastward in the direction of a mighty jungle tree just bursting into a myriad of berries. They were sulphur-breasted toucans, and they were silent, heralded only by the sound of their wings and the crash of their pigments. I can think of no other assemblage of jungle creatures more fitted to impress one with the prodigality of tropical nature. Four years before, we set ourselves to work to discover the first eggs and young of toucans, and after weeks of heartbreaking labor and disappointments we succeeded. Out of the five species of toucans living in this part of Guiana we found the nests of four, and the one which eluded us was the big sulphur-breasted fellow. I remembered so vividly the painstaking care with which, week after week, we and our Indians tramped the jungle for miles,—through swamps and over rolling hills,—at last having to admit failure; and now I sat and watched thirty, forty, fifty of the splendid birds whirr past. As the last of the fifty-four flew on to their feast of berries, I recalled with difficulty my faded visions of northern birds.

And so ended, as in the great finale of a pyrotechnic display, my two hours on a hillside clearing. I can neither enliven it with a startling escape, nor add a thrill of danger, without using as many "ifs" as would be needed to make a Jersey meadow untenable. For example, *if* I had fallen over backwards and been powerless to rise or move, I should have been killed within half an hour, for a stray column of army ants was passing within a yard of me, and death would await any helpless being falling across their path. But by searching out a copperhead and imitating Cleopatra, or with patience and persistence devouring every toadstool, the same result could be achieved in our home-town orchard. When on the march, the army ants are as innocuous at two inches as at two miles. Had I sat where I was for days and for nights, my chief danger would have been demise from sheer chagrin at my inability to grasp the deeper significance of life and its earthly activities.

III

THE HOME TOWN OF THE ARMY ANTS

FROM uniform to civilian clothes is a change transcending mere alteration of stuffs and buttons. It is scarcely less sweeping than the shift from civilian clothes to bathing-suit, which so often compels us to concentrate on remembered mental attributes, to avoid demanding a renewed introduction to estranged personality. In the home life of the average soldier, the relaxation from sustained tension and conscious routine results in a gentleness and quietness of mood for which warrior nations are especially remembered.

Army ants have no insignia to lay aside, and their swords are too firmly hafted in their own beings to be hung up as post-bellum mural decorations, or—as is done only in poster-land—metamorphosed into pruning-hooks and plowshares.

I sat at my laboratory table at Kartabo, and looked down river to the pink roof of Kalacoon, and my mind went back to the shambles of Pit

Number Five.[1] I was wondering whether I should ever see the army ants in any guise other than that of scouting, battling searchers for living prey, when a voice of the jungle seemed to hear my unexpressed wish. The sharp, high notes of white-fronted antbirds—those white-crested watchers of the ants—came to my ears, and I left my table and followed up the sound. Physically, I merely walked around the bungalow and approached the edge of the jungle at a point where we had erected a small outhouse a day or two before. But this two hundred feet might just as well have been a single step through quicksilver, hand in hand with Alice, for it took me from a world of hyoids and syrinxes, of vials and lenses and clean-smelling xylol, to the home of the army ants.

The antbirds were chirping and hopping about on the very edge of the jungle, but I did not have to go that far. As I passed the doorless entrance of the outhouse I looked up, and there was an immense mass of some strange material suspended in the upper corner. It looked like stringy, chocolate-colored tow, studded with hundreds of tiny ivory buttons. I came closer and looked

[1] See *Jungle Peace*, p. 211.

carefully at this mushroom growth which had appeared in a single night, and it was then that my eyes began to perceive and my mind to record, things that my reason besought me to reject. Such phenomena were all right in a dream, or one might imagine them and tell them to children on one's knee, with wind in the eaves—wild tales to be laughed at and forgotten. But this was daylight and I was a scientist; my eyes were in excellent order, and my mind rested after a dreamless sleep; so I had to record what I saw in that little outhouse.

This chocolate-colored mass with its myriad ivory dots was the home, the nest, the hearth, the nursery, the bridal suite, the kitchen, the bed and board of the army ants. It was the focus of all the lines and files which ravaged the jungle for food, of the battalions which attacked every living creature in their path, of the unnumbered rank and file which made them known to every Indian, to every inhabitant of these vast jungles.

Louis Quatorze once said, *"L'Etat, c'est moi!"* but this figure of speech becomes an empty, meaningless phrase beside what an army ant could boast,—*"La maison, c'est moi!"* Every rafter, beam, stringer, window-frame and doorframe, hall-way, room, ceiling, wall and floor, foundation, superstructure and roof, all were ants—living ants, distorted by stress, crowded into the dense walls, spread out to widest stretch across tie-spaces. I had thought it marvelous when I saw them arrange themselves as bridges, walks, hand-rails, buttresses, and sign-boards along the columns; but this new absorption of environment, this usurpation of wood and stone, this insinuation of themselves into the province of the inorganic world, was almost too astounding to credit.

All along the upper rim the sustaining structure was more distinctly visible than elsewhere. Here was a maze of taut brown threads stretching in places across a span of six inches, with here and there a tiny knot. These were actually tie-strings of living ants, their legs stretched almost to the breaking-point, their bodies the inconspicuous knots or nodes. Even at rest and at home, the army ants are always prepared, for every quiescent individual in the swarm was standing as erect as possible, with jaws widespread and ready, whether the great curved mahogany scimitars of the soldiers, or the little black daggers of the smaller workers. And with no eyelids to close, and eyes which were themselves a mockery, the nerve shriveling and never reaching the brain, what could sleep mean to them? Wrapped ever in an impenetrable cloak of darkness and silence, life was yet one great activity, directed, ordered, commanded by scent and odor alone. Hour after hour, as I sat close to the nest, I was aware of this odor, sometimes subtle, again wafted in strong successive waves. It was musty, like something sweet which had begun to mold; not unpleasant, but very difficult to describe; and in vain I strove to realize the importance of this faint essence—taking the place of sound, of language, of color, of motion, of form.

I recovered quickly from my first rapt realization, for a dozen ants had lost no time in ascending my shoes, and, as if at a preconcerted signal, all simultaneously sank their jaws into my person. Thus strongly recalled to the realities of life, I realized the opportunity that was offered and planned for my observation. No living thing could long remain motionless within the sphere of influence of these six-legged Boches, and yet I intended to spend days in close proximity. There was no place to hang a hammock, no over-hanging tree from which I might suspend myself spider-wise. So I sent Sam for an ordinary chair, four tin cans, and a bottle of disinfectant. I filled the tins with the tarry fluid, and in four carefully timed rushes I placed the tins in a chair-leg square. The fifth time I put the chair in place beneath the nest, but I had misjudged my distances and had to retreat with only two tins in place. Another effort, with Spartan-like disregard of the fiery bites, and my haven was ready. I hung a bag of vials, notebook, and lens on the chairback, and, with a final rush, climbed on the seat and curled up as comfortably as possible.

All around the tins, swarming to the very edge of the liquid, were the angry hosts. Close to my face were the lines ascending and descending, while just above me were hundreds of thousands, a bushel-basket of army ants, with only the strength of their thread-like legs as suspension cables. It took some time to get used to my environment, and from first to last I was never wholly relaxed, or quite unconscious of what would happen if a chair-leg broke, or a bamboo fell across the outhouse.

I swiveled round on the chair-seat and counted

eight lines of army ants on the ground, converging to the post at my elbow. Each was four or five ranks wide, and the eight lines occasionally divided or coalesced, like a nexus of capillaries. There was a wide expanse of sand and clay, and no apparent reason why the various lines of foragers should not approach the nest in a single large column. The dividing and redividing showed well how completely free were the columns from any individual dominance. There was no control by specific individuals or soldiers, but, the general route once established, the governing factor was the odor of contact.

The law to pass where others have passed is immutable, but freedom of action or individual desire dies with the malleable, plastic ends of the foraging columns. Again and again came to mind the comparison of the entire colony or army with a single organism; and now the home, the nesting swarm, the focus of central control, seemed like the body of this strange amorphous organism—housing the spirit of the army. One thinks of a column of foragers as a tendril with only the tip sensitive and growing and moving, while the corpuscle-like individual ants are driven in the current of blind instinct to and fro, on their chemical errands. And then this whole theory, this most vivid simile, is quite upset by the sights that I watch in the suburbs of this ant home!

The columns were most excellent barometers, and their reaction to passing showers was invariable. The clay surface held water, and after each downfall the pools would be higher, and the contour of the little region altered. At the first few drops, all the ants would hasten, the throbbing corpuscles speeding up. Then, as the rain came down heavier, the column melted away, those near each end hurrying to shelter and those in the center crawling beneath fallen leaves and bits of clod and sticks. A moment before, hundreds of ants were trudging around a tiny pool, the water lined with ant handrails, and in shallow places, veritable formicine pontoons,—large ants which stood up to their bodies in water, with the booty-laden host passing over them. Now, all had vanished, leaving only a bare expanse of splashing drops and wet clay. The sun broke through and the residue rain tinkled from the bamboos.

As gradually as the growth of the rainbow above the jungle, the lines reformed themselves.

Scouts crept from the jungle-edge at one side, and from the post at my end, and felt their way, fan-wise, over the rain-scoured surface; for the odor, which was both sight and sound to these ants, had been washed away—a more serious handicap than mere change in contour. Swiftly the wandering individuals found their bearings again. There was deep water where dry land had been, but, as if by long-planned study of the work of sappers and engineers, new pontoon bridges were thrown across, washouts filled in, new cliffs explored, and easy grades established; and by the time the bamboos ceased their own private after-shower, the columns were again running smoothly, battalions of eager light infantry hastening out to battle, and equal hosts of loot-laden warriors hurrying toward the home nest. Four minutes was the average time taken to reform a column across the ten feet of open clay, with all the road-making and engineering feats which I have mentioned, on the part of ants who had never been over this new route before.

Leaning forward within a few inches of the post, I lost all sense of proportion, forgot my awkward human size, and with a new perspective became an equal of the ants, looking on, watching every passer-by with interest, straining with the bearers of the heavy loads, and breathing more easily when the last obstacle was overcome and home attained. For a period I plucked out every bit of good-sized booty and found that almost all were portions of scorpions from far-distant dead logs in the jungle, creatures whose strength and poisonous stings availed nothing against the attacks of these fierce ants. The loads were adjusted equably, the larger pieces carried by the big, white-headed workers, while the smaller ants transported small eggs and larvæ. Often, when a great mandibled soldier had hold of some insect, he would have five or six tiny workers surrounding him, each grasping any projecting part of the loot, as if they did not trust him in this menial capacity,—as an anxious mother would watch with doubtful confidence a big policeman wheeling her baby across a crowded street. These workers were often diminutive Marcelines, hindering rather than aiding in the progress. But in every phase of activity of these ants there was not an ounce of intentionally lost power, or a moment of time wilfully gone to waste. What a commentary on Bolshevism!

Now that I had the opportunity of quietly watching the long, hurrying columns, I came hour by hour to feel a greater intimacy, a deeper enthusiasm for their vigor of existence, their unfailing life at the highest point of possibility of achievement. In every direction my former desultory observations were discounted by still greater accomplishments. Elsewhere I have recorded the average speed as two and a half feet in ten seconds, estimating this as a mile in three and a half hours. An observant colonel in the American army has laid bare my congenitally hopeless mathematical inaccuracy, and corrected this to five hours and fifty-two seconds. Now, however, I established a wholly new record for the straight-away dash for home of the army ants. With the handicap of gravity pulling them down, the ants, both laden and unburdened, averaged ten feet in twenty seconds, as they raced up the post. I have now called in an artist and an astronomer to verify my results, these two being the only living beings within hailing distance as I write, except a baby red howling monkey curled up in my lap, and a toucan, sloth, and green boa, beyond my laboratory table. Our results are identical, and I can safely announce that the amateur record for speed of army ants is equivalent to a mile in two hours and fifty-six seconds; and this when handicapped by gravity and burdens of food, but with the incentive of approaching the end of their long journey.

As once before, I accidentally disabled a big worker that I was robbing of his load, and his entire abdomen rolled down a slope and disappeared. Hours later in the afternoon, I was summoned to view the same soldier, unconcernedly making his way along an outward-bound column, guarding it as carefully as if he had not lost the major part of his anatomy. His mandibles were ready, and the only difference that I could see was that he could make better speed than others of his caste. That night he joined the general assemblage of cripples quietly awaiting death, halfway up to the nest.

I know of no highway in the world which surpasses that of a big column of army ants in exciting happenings, although I usually had the feeling which inspired Kim as he watched the Great White Road, of understanding so little of all that was going on. Early in the morning there were only outgoing hosts; but soon eddies were seen in the swift current, vortexes made by a single ant here and there forcing its way against the stream. Unlike penguins and human beings, army ants have no rule of the road as to right and left, and there is no lessening of pace or turning aside for a heavily laden drogher. Their blindness caused them to bump squarely into every individual, often sending load and carrier tumbling to the bottom of a vertical path. Another constant loss of energy was a large cockroach leg, or scorpion segment, carried by several ants. Their insistence on trying to carry everything beneath their bodies caused all sorts of comical mishaps. When such a large piece of booty appeared, it was too much of a temptation, and a dozen outgoing ants would rush up and seize hold for a moment, the consequent pulling in all directions reducing progress at once to zero.

Until late afternoon few ants returned without carrying their bit. The exceptions were the cripples, which were numerous and very pitiful. From such fierce strenuousness, such virile activity, as unending as elemental processes, it seemed a very terrible drop to disability, to the utilizing of every atom of remaining strength to return to the temporary home nest—that instinct which drives so many creatures to the same homing, at the approach of death.

Even in their helplessness they were wonderful. To see a big black-headed worker struggling up a post with five short stumps and only one good hind leg, was a lesson in achieving the impossible. I have never seen even a suspicion of aid given to any cripple, no matter how slight or how complete the disability; but frequently a strange thing occurred, which I have often noticed but can never explain. One army ant would carry another, perhaps of its own size and caste, just as if it were a bit of dead provender; and I always wondered if cannibalism was to be added to their habits. I would capture both, and the minute they were in the vial, the dead ant would come to life, and with equal vigor and fury both would rush about their prison, seeking to escape, becoming indistinguishable in the twinkling of an eye.

Very rarely an ant stopped and attempted to clean another which had become partly disabled through an accumulation of gummy sap or other encumbering substance. But when a leg or other organ was broken or missing, the odor of the ant-blood seemed to arouse only suspicion and to banish sympathy, and after a few casual wavings of antennæ, all passed by on the other side. Not only this, but the unfortunates were actually in danger of attack within the very lines of traffic of the legionaries. Several times I noticed small rove-beetles accompanying the ants, who paid little attention to them. Whenever an ant became suspicious and approached with a raised-eyebrow gesture of antennæ, the beetles turned their backs quickly and raised threatening tails. But I did not suspect the vampire or thug-like character of these guests—tolerated where any other insect would have been torn to pieces at once. A large crippled worker, hobbling along, had slipped a little away from the main line, when I was astonished to see two rove-beetles rush at him and bite him viciously, a third coming up at once and joining in. The poor worker had no possible chance against this combination, and he went down after a short, futile struggle. Two small army ants now happened to pass, and after a preliminary whiffing with waving antennæ, rushed joyously into the *mêlée*. The beetles had a cowardly weapon, and raising their tails, ejected a drop or two of liquid, utterly confusing the ants, which turned and hastened back to the column. For the next few minutes, until the scent wore off, they aroused suspicion wherever they went. Meanwhile, the hyena-like rove-beetles, having hedged themselves within a barricade of their malodor, proceeded to feast, quarreling with one another as such cowards are wont to do.

Thus I thought, having identified myself with the army ants. From a broader, less biased point of view, I realized that credit should be given to the rove-beetles for having established themselves in a zone of such constant danger, and for being able to live and thrive in it.

The columns converged at the foot of the post, and up its surface ran the main artery of the nest. Halfway up, a flat board projected, and here the column divided for the last time, half going on directly into the nest, and the other half turning aside, skirting the board, ascending a bit of perpendicular canvas, and entering the nest from the rear. The entrance was well guarded by a veritable moat and drawbridge of living ants. A foot away, a flat mat of ants, mandibles outward, was spread, over which every passing individual stepped. Six inches farther, and the sides of the mat thickened, and in the last three inches these sides met overhead, forming a short tunnel at the end of which the nest began.

And here I noticed an interesting thing. Into this organic moat or tunnel, this living mouth of an inferno, passed all the booty-laden foragers, or those who for some reason had returned empty-mouthed. But the outgoing host seeped gradually from the outermost nest-layer—a gradual but fundamental circulation, like that of ocean currents. Scorpions, eggs, caterpillars, glass-like wasp pupæ, roaches, spiders, crickets, —all were drawn into the nest by a maelstrom of hunger, funneling into the narrow tunnel; while from over all the surface of the swarm there crept forth layer after layer of invigorated, implacable seekers after food.

The mass of ants composing the nest appeared so loosely connected that it seemed as if a touch would tear a hole, a light wind rend the supports. It was suspended in the upper corner of the doorway, rounded on the free sides, and measured roughly two feet in diameter—an unnumbered host of ants. Those on the surface were in very slow but constant motion, with legs shifting and antennæ waving continually. This quivering on the surface of the swarm gave it the appearance of the fur of some terrible animal—fur blowing in the wind from some unknown, deadly desert. Yet so cohesive was the entire mass, that I sat close beneath it for the best part of two days and not more than a dozen ants fell upon me. There was, however, a constant rain of egg-cases and pupa-skins and the remains of scorpions and grasshoppers, the residue of the booty which was being poured in. These wrappings and inedible casing were all brought to the surface and dropped. This was reasonable, but what I could not comprehend was a constant falling of small living larvæ. How anything except army ants could emerge alive from such a sinister swarm was inconceivable. It took some resolution to stand up under the nest, with my face only a foot away from this slowly seething mass of widespread jaws. But I had to discover where the falling larvæ came from, and after a time I found that they were immature army ants. Here and there a small worker would appear, carrying in its mandibles a young larva; and while most made their way through the maze of mural legs and bodies and ultimately disappeared again, once in a while the burden was dropped and fell to the floor of the outhouse. I can account for this only by presuming that a certain percentage of the nurses were very young and inexperienced workers and dropped their burdens inadvertently. There was certainly no intentional casting out of these offspring, as was so obviously the case with the débris from the food of the colony. The eleven or twelve ants which fell upon me during my watch were all smaller workers, no larger ones losing their grip.

While recording some of these facts, I dropped my pencil, and it was fully ten minutes before the black mass of enraged insects cleared away, and I could pick it up. Leaning far over to secure it, I was surprised by the cleanliness of the floor around my chair. My clothes and notepaper had been covered with loose wings, dry skeletons of insects and the other débris, while hundreds of other fragments had sifted down past me. Yet now that I looked seeingly, the whole area was perfectly clean. I had to assume a perfect jack-knife pose to get my face near enough to the floor; but, achieving it, I found about five hundred ants serving as a street-cleaning squad. They roamed aimlessly about over the whole floor, ready at once to attack anything of mine, or any part of my anatomy which might come close enough, but otherwise stimulated to activity only when they came across a bit of rubbish from the nest high overhead. This was at once seized and carried off to one of two neat piles in far corners. Before night these kitchen middens were an inch or two deep and nearly a foot in length, composed, literally, of thousands of skins, wings, and insect armor. There was not a scrap of dirt of any kind which had not been gathered into one of the two piles. The nest was nine feet above the floor, a distance (magnifying ant height to our own) of nearly a mile, and yet the care lavished on the cleanliness of the earth so far below was as thorough and well done as the actual provisioning of the colony.

As I watched the columns and the swarm-nest hour after hour, several things impressed me;—the absolute silence in which the ants worked;—such ceaseless activity without sound one associates only with a cinema film; all around me was tremendous energy, marvelous feats of achievement, super-human instincts, the ceaseless movement of tens of thousands of legionaries; yet no tramp of feet, no shouts, no curses, no welcomes, no chanties. It was uncanny to think of a race of creatures such as these, dreaded by every living being, wholly dominant in their continent-wide sphere of action, yet born, living out their lives, and dying, dumb and blind, with no possibility of comment on life and its fullness, of censure or of applause.

The sweeping squad on the floor was interesting because of its limited field of work at such a distance from the nest; but close to my chair were a number of other specialized zones of activity, any one of which would have afforded a fertile field for concentrated study. Beneath the swarm on the white canvas, I noticed two large spots of dirt and moisture, where very small flies were collected. An examination showed that this was a second, nearer dumping-ground for all the garbage and refuse of the swarm which could not be thrown down on the kitchen middens far below. And here were tiny flies and other insects acting as scavengers, just as the hosts of vultures gather about the slaughter-house of Georgetown.

The most interesting of all the phases of life of the ants' home town, were those on the horizontal board which projected from the beam and stretched for several feet to one side of the swarm. This platform was almost on a level with my eyes, and by leaning slightly forward on the chair, I was as close as I dared go. Here many ants came from the incoming columns, and others were constantly arriving from the nest itself. It was here that I realized my good fortune and the achievement of my desires, when I first saw an army ant at rest. One of the first arrivals after I had squatted to my post, was a big soldier with a heavy load of roach meat. Instead of keeping on straight up the post, he turned abruptly and dropped his load. It was instantly picked up by two smaller workers and carried on and upward toward the nest. Two other big fellows arrived in quick succession, one with a load which he relinquished to a drogher-in-waiting. Then the three weary warriors stretched their legs one after another and commenced to clean their antennæ. This lasted only for a moment, for three or four tiny ants rushed at each of the larger ones and began as thorough a cleaning as masseurs or Turkish-bath attendants. The three arrivals were at once hustled away to a distant part of the board and there cleaned from end to end. I found that the focal length of my 8-diameter lens was just out of reach of the ants, so I focused carefully on one of the soldiers and watched the entire process. The small ants scrubbed and scraped him with their jaws, licking him and removing every particle of dirt. One even crawled under him and worked away at his upper leg-joints, for all the world as a mechanic will creep under a car. Finally, I was delighted to see him do what no car ever does, turn completely over and lie quietly on his back with his legs in air, while his diminutive helpers overran him and gradually got him into shape for future battles and foraging expeditions.

On this resting-stage, within well-defined limits, were dozens of groups of two cleaning one another, and less numerous parties of the tiny professionals working their hearts out on battle-worn soldiers. It became more and more apparent that in the creed of the army ants, cleanliness comes next to military effectiveness.

Here and there I saw independent individuals cleaning themselves and going through the most un-ant-like movements. They scraped their jaws along the board, pushing forward like a dog trying to get rid of his muzzle; then they turned on one side and passed the opposite legs again and again through the mandibles; while the last performance was to turn over on their backs and roll from side to side, exactly as a horse or donkey loves to do.

One ant, I remember, seemed to have something seriously wrong. It sat up on its bent-under abdomen in a most comical fashion, and was the object of solicitude of every passing ant. Sometimes there were thirty in a dense group, pushing and jostling; and, like most of our city crowds, many seemed to stop only long enough to have a moment's morbid sight, or to ask some silly question as to the trouble, then to hurry on. Others remained, and licked and twiddled him with their antennæ for a long time. He was in this position for at least twenty minutes. My curiosity was so aroused that I gathered him up in a vial, whereat he became wildly excited and promptly regained full use of his legs and faculties. Later, when I examined him under the lens, I could find nothing whatever wrong.

Off at one side of the general cleaning and reconstruction areas was a pitiful assemblage of cripples which had had enough energy to crawl back, but which did not attempt, or were not allowed, to enter the nest proper. Some had one or two legs gone, others had lost an antenna or had an injured body. They seemed not to know what to do—wandering around, now and then giving one another a half-hearted lick. In the midst was one which had died, and two others, each badly injured, were trying to tug the body along to the edge of the board. This they succeeded in doing after a long series of efforts, and down and down fell the dead ant. It was promptly picked up by several kitchen-middenites and unceremoniously thrown on the pile of nest-débris. A load of booty had been dumped among the cripples, and as each wandered close to it, he seemed to regain strength for a moment, picked up the load, and then dropped it. The sight of that which symbolized almost all their life-activity aroused them to a momentary forgetfulness of their disabilities. There was no longer any place for them in the home or in the columns of the legionaries. They had been court-martialed under the most implacable, the most impartial law in the world—the survival of the fit, the elimination of the unfit.

The time came when we had to get at our stored supplies, over which the army ants were such an effective guard. I experimented on a running column with a spray of ammonia and found that it created merely temporary inconvenience, the ants running back and forming a new trail. Formaline was more effective, so I sprayed the nest-swarm with a fifty-per-cent solution, strong enough, one would think, to harden the very boards. It certainly created a terrible commotion, and strings of the ants, two feet long, hung dangling from the nest. The heart of the colony came into view, with thousands of eggs and larvæ, looking like heaps of white rice-grains. Every ant seized one or the other and sought escape by the nearest way, while the soldiers still defied the world. The gradual disintegration revealed an interior meshed like a wasp's nest, chambered and honeycombed with living tubes and walls. Little by little the taut guy-ropes, lathes, braces, joists, all sagged and melted together, each cell-wall becoming dynamic, now expanding, now contracting; the ceilings vibrant with waving legs, the floors a seething mass of jaws and antennæ. By the time it was dark, the swarm was dropping in sections to the floor.

On the following morning new surprises awaited me. The great mass of the ants had moved in the night, vanishing with every egg and immature larva; but there was left in the corner of the flat board a swarm of about one-quarter of the entire number, enshrouding a host of older larvæ. The cleaning zones, the cripples' gathering-room, all had given way to new activities, on the flat board, down near the kitchen middens, and in every horizontal crack.

The cause of all this strange excitement, this braving of the terrible dangers of fumes which had threatened to destroy the entire colony the night before, suddenly was made plain as I watched. A critical time was at hand in the lives of the all-precious larvæ, when they could not be moved—the period of spinning, of beginning the transformation from larvæ to pupæ. This evidently was an operation which had to take place outside the nest and demanded some sort of light covering. On the flat board were several thousand ants and a dozen or more groups of full-grown larvæ. Workers of all sizes were searching everywhere for some covering for the tender immature creatures. They had chewed up all available loose splinters of wood, and near the rotten, termite-eaten ends, the sound of dozens of jaws gnawing all at once was plainly audible. This unaccustomed, unmilitary labor produced a quantity of fine sawdust, which was sprinkled over the larvæ. I had made a partition of a bit of a British officer's tent which I had used in India and China, made of several layers of colored canvas and cloth. The ants found a loose end of this, teased it out and unraveled it, so that all the larvæ near by were blanketed with a gay, parti-colored covering of fuzz.

All this strange work was hurried and carried on under great excitement. The scores of big soldiers on guard appeared rather ill at ease, as if they had wandered by mistake into the wrong department. They sauntered about, bumped into larvæ, turned and fled. A constant stream of workers from the nest brought hundreds more larvæ; and no sooner had they been planted and débris of sorts sifted over them, than they began spinning. A few had already swathed themselves in cocoons—exceedingly thin coverings of pinkish silk. As this took place out of the nest,—in the jungle they must be covered with wood and leaves. The vital necessity for this was not apparent, for none of this débris was incorporated into the silk of the cocoons, which were clean and homogeneous. Yet the hundreds of ants gnawed and tore and labored to gather this little dust, as if their very lives depended upon it.

With my hand-lens focused just beyond mandible reach of the biggest soldier, I leaned forward from my insulated chair, hovering like a great astral eye looking down at this marvelously important business of little lives. Here were thousands of army ants, not killing, not carrying booty, nor even suspended quiescent as organic molecules in the structure of the home, yet in feverish activity equaled only by battle, making ready for the great change of their foster offspring. I watched the very first thread of silk drawn between the larva and the outside world, and in an incredibly short time the cocoon was outlined in a tissue-thin, transparent aura, within which the tenant could be seen skilfully weaving its own shroud.

When first brought from the nest, the larvæ lay quite straight and still; but almost at once they bent far over in the spinning position. Then some officious worker would come along, and the unfortunate larva would be snatched up, carried off, and jammed down in some neighboring empty space, like a bolt of cloth rearranged upon a shelf. Then another ant would approach, antennæ the larva, disapprove, and again shift its position. It was a real survival of the lucky, as to who should avoid being exhausted by kindness and over-solicitude. I uttered many a chuckle at the half-ensilked unfortunates being toted about like mummies, and occasionally giving a sturdy, impatient kick which upset their tormentors and for a moment created a little swirl of mild excitement.

There was no order of packing. The larvæ were fitted together anyway, and meagerly covered with dust of wood and shreds of cloth. One big tissue of wood nearly an inch square was too great a temptation to be let alone, and during the course of my observation it covered in turn almost every group of larvæ in sight, ending by being accidentally shunted over the edge and killing a worker near the kitchen middens. There was only a single layer of larvæ; in no case were they piled up, and when the platform became crowded, a new column was formed and hundreds taken outside. To the casual eye there was no difference between these legionaries and a column bringing in booty of insects, eggs, and pupæ; yet here all was solicitude, never a bite too severe, or a blunder of undue force.

The sights I saw in this second day's accessible nest-swarm would warrant a season's meditation and study, but one thing impressed me above all others. Sometimes, when I carefully pried open one section and looked deep within, I could see large chambers with the larvæ in piles, besides being held in the mandibles of the components of the walls and ceilings. Now and then a curious little ghost-like form would flit across the chamber, coming to rest, gnome-like, on larva or ant. Again and again I saw these little springtails skip through the very scimitar mandibles of a soldier, while the workers paid no attention to them. I wondered if they were not quite odorless, intangible to the ants, invisible guests which lived close to them, going where, doing what they willed, yet never perceived by the thousands of inhabitants. They seemed to live in a kind of fourth dimensional state, a realm comparable to that which we people with ghosts and spirits. It was a most uncanny, altogether absorbing, intensely interesting relationship; and sometimes, when I ponder on some general aspect of the great jungle,—a forest of greenheart, a mighty rushing river, a crashing, blasting thunderstorm,—my mind suddenly reverts by way of contrast to the tiny ghosts of springtails flitting silently among the terrible living chambers of the army ants.

On the following morning I expected to achieve still greater intimacy in the lives of the mummy soldier embryos; but at dawn every trace of nesting swarm, larvæ, pupæ and soldiers was gone. A few dead workers were being already carried off by small ants which never would have dared approach them in life. A big blue morpho butterfly flapped slowly past out of the jungle, and in its wake came the distant notes—high and sharp—of the white-fronted antbirds; and I knew that the legionaries were again abroad, radiating on their silent, dynamic paths of life from some new temporary nest deep in the jungle.

IV

A JUNGLE BEACH

A JUNGLE moon first showed me my beach. For a week I had looked at it in blazing sunlight, walked across it, even sat on it in the intervals of getting wonted to the new laboratory; yet I had not perceived it. Colonel Roosevelt once said to me that he would rather perceive things from the point of view of a field-mouse, than be a human being and merely see them. And in my case it was when I could no longer see the beach that I began to discern its significance.

This British Guiana beach, just in front of my Kartabo bungalow, was remarkably diversified, and in a few steps, or strokes of a paddle, I could pass from clean sand to mangroves and muckamucka swamp, thence to out-jutting rocks, and on to the Edge of the World, all within a distance of a hundred yards. For a time my beach walks resulted in inarticulate reaction. After months in the blindfolded canyons of New York's streets, a hemicircle of horizon, a hemisphere of sky, and a vast expanse of open water lent itself neither to calm appraisal nor to impromptu cuff-notes.

It was recalled to my mind that the miracle of sunrise occurred every morning, and was not a rather belated alternation of illumination, following the quenching of Broadway's lights. And the moon I found was as dependable as when I timed my Himalayan expeditions by her shadowings. To these phenomena I soon became re-accustomed, and could watch a bird or outwit an insect in the face of a foreglow and silent burst of flame that shamed all the barrages ever laid down. But cosmic happenings kept drawing my attention and paralyzing my activities for long afterward. With a double rainbow and four storms in action at once; or a wall of rain like sawn steel slowly drawing up one river while the Mazaruni remains in full sunlight; with Pegasus galloping toward the zenith at midnight and the Pleiades just clearing the Penal Settlement, I could not always keep on dissecting, or recording, or verifying the erroneousness of one of my recently formed theories.

There was Thuban, gazing steadily upon my little mahogany bungalow, as, six millenniums ago, he had shone unfalteringly down the little stone tube that led his rays into the Queen's Chamber, in the very heart of great Cheops. Just clearing a low palm was the present North Star, while, high above, Vega shone, patiently waiting to take her place half a million years hence. When beginning her nightly climb, Vega drew a thin, trembling thread of argent over the still water, just as in other years she had laid for me a slender silver strand of wire across frozen snow, and on one memorable night traced the ghost of a reflection over damp sand near the Nile—pale as the wraiths of the early Pharaohs.

Low on the eastern horizon, straight outward from my beach, was the beginning and end of the great zodiac band—the golden Hamal of Aries and the paired stars of Pisces; and behind, over the black jungle, glowed the Southern Cross. But night after night, as I watched on the beach, the sight which moved me most was the dull speck of emerald mist, a merest smudge on the slate of the heavens,—the spiral nebula in Andromeda,—a universe in the making, of a size unthinkable to human minds.

The power of my jungle beach to attract and

hold attention was not only direct and sensory, —through sight and sound and scent,—but often indirect, seemingly by occult means. Time after time, on an impulse, I followed some casual line of thought and action, and found myself at last on or near the beach, on a lead that eventually would take me to the verge or into the water.

Once I did what for me was a most unusual thing. I woke in the middle of the night without apparent reason. The moonlight was pouring in a white flood through the bamboos, and the jungle was breathless and silent. Through my window I could see Jennie, our pet monkey, lying aloft, asleep on her little verandah, head cushioned on both hands, tail curled around her dangling chain, as a spider guards her web-strands for hint of disturbing vibrations. I knew that the slightest touch on that chain would awaken her, and indeed it seemed as if the very thought of it had been enough; for she opened her eyes, sent me the highest of insect-like notes and turned over, pushing her head within the shadow of her little house. I wondered if animals, too, were, like the Malays and so many savage tribes, afraid of the moonlight—the "luna-cy" danger in those strange color-strained rays, whose power must be greater than we realize. Beyond the monkey roosted Robert, the great macaw, wide-awake, watching me with all that broadside of intensive gaze of which only a parrot is capable.

The three of us seemed to be the only living things in the world, and for a long time we—monkey, macaw, and man—listened. Then all but the man became uneasy. The monkey raised herself and listened, uncurled her tail, shifted, and listened. The macaw drew himself up, feathers close, forgot me, and listened. They, unlike me, were not merely listening—they were hearing something. Then there came, very slowly and deliberately, as if reluctant to break through the silent moonlight, a sound, low and constant, impossible to identify, but clearly audible even to my ears. For just an instant longer it held, sustained and quivering, then swiftly rose into a crashing roar—the sound of a great tree falling. I sat up and heard the whole long descent; but at the end, after the moment of silence, there was no deep boom—the sound of the mighty bole striking and rebounding from the earth itself. I wondered about this for a while; then the monkey and I went to sleep, leaving the macaw alone conscious in the moonlight, watching through the night with his great round, yellow orbs, and thinking the thoughts that macaws always think in the moonlight.

The next day the macaw and the monkey had forgotten all about the midnight sound, but I searched and found why there was no final boom. And my search ended at my beach. A bit of overhanging bank had given way and a tall tree had fallen headlong into the water, its roots sprawling helplessly in mid-air. Like rats deserting a sinking ship, a whole Noah's ark of tree-living creatures was hastening along a single cable shorewards: tree-crickets; ants laden with eggs and larvæ; mantids gesticulating as they walked, like old men who mumble to themselves; woodroaches, some green and leaf-like, others, facsimiles of trilobites—but fleet of foot and with one goal.

What was a catastrophe for a tree and a shift of home for the tenants was good fortune for me, and I walked easily out along the trunk and branches and examined the strange parasitic growths and the homes which were being so rapidly deserted. The tide came up and covered the lower half of the prostrate tree, drowning what creatures had not made their escape and quickening the air-plants with a false rain, which in course of time would rot their very hearts.

But the first few days were only the overture of changes in this shift of conditions. Tropic vegetation is so tenacious of life that it struggles and adapts itself with all the cunning of a Japanese wrestler. We cut saplings and thrust them into mud or the crevices of rocks at low tide far from shore, to mark our channel, and before long we have buoys of foliage banners waving from the bare poles above water. We erect a tall bamboo flagpole on the bank, and before long our flag is almost hidden by the sprouting leaves, and the pulley so blocked that we have occasionally to lower and lop it.

So the fallen tree, still gripping the nutritious bank with a moiety of roots, turned slowly in its fibrous stiffness and directed its life and sap and hopes upward. During the succeeding weeks I watched trunk and branches swell and bud out new trunks, new branches, guided, controlled, by gravity, light, and warmth; and just beyond the reach of the tides, leaves sprouted, flowers opened and fruit ripened. Weeks after the last slow invertebrate plodder had made his escape shorewards, the taut liana strand was again crowded with a mass of passing life—a maze of vines and creepers, whose tendrils and suckers reached and curled and pressed onward, fighting for gangway to shore, through days and weeks, as the animal life which preceded them had made the most of seconds and minutes.

The half-circle of exposed raw bank became in its turn the center of a myriad activities. Great green kingfishers began at once to burrow; tiny emerald ones chose softer places up among the wreckage of wrenched roots; wasps came and chopped out bits for the walls and partitions of their cells; spiders hung their cobwebs between ratlines of rootlets; and hummingbirds promptly followed and plucked them from their silken nets, and then took the nets to bind their own tiny air-castles. Finally, other interests intervened, and like Jennie and Robert, I gradually forgot the tree that fell without an echo.

In the jungle no action or organism is separate, or quite apart, and this thing which came to the three of us suddenly at midnight led by devious means to another magic phase of the shore.

A little to the south along my beach is the Edge of the World. At least, it looks very much as I have always imagined that place must look, and I have never been beyond it; so that, after listening to many arguments in courts of law, and hearing the reasoning of bolsheviki, teetotalers, and pacifists, I feel that I am quite reasonable as human beings go. And best of all, it hurts no one, and annoys only a few of my scientific friends, who feel that one cannot indulge in such ideas at the wonderful hour of twilight, and yet at eight o'clock the following morning describe with impeccable accuracy the bronchial semi-rings, and the intricate mosaic of cartilage which characterizes and supports the *membranis tympaniformis* of *Attila thamnophiloides*; a dogma which halves life and its interests.

The Edge of the World has always meant a place where usual things are different; and my southern stretch of beach was that, because of roots. Whenever in digging I have come across a root and seen its living flesh, perhaps pink or rose or pale green, so far underground, I have desired to know roots better; and now I found my opportunity. I walked along the proper traii, through right and usual trees, with reasonable foliage and normal trunks, and suddenly I stepped down over the Edge. Overhead and all around there was still the foliage. It shut out the sun except for greenish, moderated spots and beams. The branches dipped low in front over the water, shutting out the sky except along the tops of the cross-river jungle. Thus a great green-roofed chamber was formed; and here, between jungle and the water-level of the world, was the Kingdom of the Roots.

Great trees had in their youth fallen far forward, undermined by the water, then slowly taken a new reach upward and stretched forth great feet and hands of roots, palms pressing against the mud, curved backs and thews of shoulders braced against one another and the drag of the tides. Little by little the old prostrate trunks were entirely obliterated by this fantastic network. There were no fine fibers or rootlets here; only great beams and buttresses, bridges and up-ended spirals, grown together or spreading wide apart. Root merged with trunk, and great boles became roots and then boles again in this unreasonable land. For here, in place of damp, black mold and soil, water alternated with dark-shadowed air; and so I was able for a time to live the life of a root, resting quietly among them, watching and feeling them, and moving very slowly, with no thought of time, as roots must.

I liked to wait until the last ripple had lapped against the sand beneath, and then slip quietly in from the margin of the jungle and perch—like a great tree-frog—on some convenient shelf. Seumas and Brigid would have enjoyed it, in spite of the fact that the Leprechauns seemed to have just gone. I found myself usually in a little room, walled with high-arched, thin sheets of living roots, some of which would form solid planks three feet wide and twelve long, and only an inch or two in thickness. These were always on edge, and might be smooth and sheer, or suddenly sprout five stubby, mittened fingers, or pairs of curved and galloping legs—and this thought gave substance to the simile which had occurred again and again: these trees reminded me of centaurs with proud, upright man torsos, and great curved backs. In one, a root dropped down and rested on the back, as a centaur who turns might rest his hand on his withers.

When I chanced upon an easy perch, and a stray idea came to mind, I squatted or sat or sprawled, and wrote, and strange things often happened to me. Once, while writing rapidly on a small sheet of paper, I found my lines growing closer and closer together until my fingers cramped, and the consciousness of the change overlaid the thoughts that were driving hand and pen. I then realized that, without thinking, I had been following a succession of faint lines, cross-ruled on my white paper, and looking up, I saw that a leaf-filtered opening had reflected strands of a spider-web just above my head, and I had been adapting my lines to the narrow spaces, my chirography controlled by cobweb shadows.

The first unreality of the roots was their rigidity. I stepped from one slender tendon of wood to the next, expecting a bending which never occurred. They might have been turned to stone, and even little twigs resting on the bark often proved to have grown fast. And this was the more unexpected because of the grace of curve and line, fold upon fold, with no sharp angles, but as full of charm of contour as their grays and olives were harmonious in color. Photographs showed a little of this; sketches revealed more; but the great splendid things themselves, devoid of similes and human imagination, were soul-satisfying in their simplicity.

I seldom sat in one spot more than a few minutes, but climbed and shifted, tried new seats, couches, perches, grips, sprawling out along the tops of two parallel monsters, or slipping under their bellies, always finding some easy way to swing up again. Two openings just permitted me to squeeze through, and I wondered whether, in another year, or ten, or fifty, the holes would have grown smaller. I became imbued with the quiet joy of these roots, so that I hated to touch the ground. Once I stepped down on the beach after something I had dropped, and the soft yielding of the sand was so unpleasant that I did not afterwards leave this strange mid-zone until I had to return. Unlike Antæus, I seemed to gain strength and poise by disassociation with the earth.

Here and there were pockets in the folds of the sweeping draperies, and each pocket was worth picking. When one tried to paint the roots, these pockets seemed made expressly to take the place of palette cups, except that now and then a crab resented the infusion of Hooker's green with his Vandyke brown puddle, and seized the end of the brush. The crabs were worthy tenants of such strange architecture, with comical eyes twiddling on the end of their stalks, and their white-mittened fists feinting and threatening as I looked into their little dark rain or tide-pools.

I found three pockets on one wall, which seemed as if they must have been "salted" for my benefit; and in them, as elsewhere on my beach, the two extremes of life met. The topmost one, curiously enough, contained a small crab, together with a large water-beetle at the farther end. Both seemed rather self-conscious, and there was no hint of fraternizing. The beetle seemed to be merely existing until darkness, when he could fly to more water and better company; and the crab appeared to be waiting for the beetle to go.

The next pocket was a long, narrow, horizontal fold, and I hoped to find real excitement among its aquatic folk; but to my surprise it had no bottom, but was a deep chute or socket, opening far below to the sand. However, this was not my discovery, and I saw dimly a weird little head looking up at me—a gecko lizard, which called this crevice home and the crabs neighbors. I hailed him as the only other backboned friend who shared the root-world with me, and then listened to a high, sweet tone, which came forth in swinging rhythm. It took some time for my eyes to become accustomed to the semi-darkness, and then I saw what the gecko saw—a big yellow-bodied fly humming in this cavern, and swinging in a small orbit as she sang. Now and then she dashed out past me and hovered in mid-air, when her note sank to a low, dull hum. Back again, and the sound rose and fell, and gained ten times in volume from the echo or reverberations. Each time she passed, the little lizard licked his chops and swallowed—a sort of vicarious expression of faith or desire; or was he in a Christian Science frame of mind, saying, "My, how good that fly tasted!" each time the dipteron passed? The fly was just as inexplicable, braving danger and darkness time after time, to leave the sunshine and vibrate in the dusk to the enormously magnified song of its wings.

With eyes that had forgotten the outside light, I leaned close to the opening and rested my forehead against the lichens of the wall of wood. The fly was frightened away, the gecko slipped lower, seemingly without effort, and in a hollowed side of the cavernous root I saw a mist, a quivering, so tenuous and indistinct that at first it might have been the dancing of motes. I saw that they were living creatures—the most delicate of tiny crane-flies—at rest looking like long-legged mosquitoes. Deep within this root, farther from the light than even the singing fly had ventured, these tiny beings whirled madly in mid-air—subterranean dervishes, using up energy for their own inexplicable ends, of which one very interested naturalist could make nothing.

Three weeks afterward I happened to pass at high tide in the canoe and peered into this pocket. The gecko was where geckos go in the space of three weeks, and the fly also had vanished, either within or without the gecko. But the crane-flies were still there: to my roughly appraising eyes the same flies, doing the same dance in exactly the same place. Three weeks later, and again I returned, this time intentionally, to see whether the dance still continued; and it was in full swing. That same night at midnight I climbed down, flashed a light upon them, and there they whirled and vibrated, silently, incredibly rapid, unceasingly.

After a thousand hours all the surroundings had changed. New leaves had sprouted, flowers faded and turned to fruit, the moon had twice attained her full brightness, our earth and sun and the whole solar system had swept headlong a full two-score million miles on the endless swing toward Vega. Only the roots and the crane-flies remained. A thousand hours had apparently made no difference to them. The roots might have been the granite near by, fashioned by primeval earth-flame, and the flies but vibrating atoms within the granite, made visible by some alchemy of elements in this weird Rim of the World.

And so a new memory is mine; and when one of these insects comes to my lamp in whatever part of the world, fluttering weakly, legs breaking off at the slightest touch, I shall cease to worry about the scientific problems that loom too great for my brain, or about the imperfection of whatever I am doing, and shall welcome the crane-fly and strive to free him from this fatal passion for flame, directing him again into the night; for he may be looking for a dark pocket in a root, a pocket on the Edge of the World, where crane-flies may vibrate with their fellows in an eternal dance. And so, in some ordained way, he will fulfil his destiny and I acquire merit.

To write of sunrises and moonlight is to commit literary harikiri; but as that terminates life, so may I end this. And I choose the morning and the midnight of the sixth of August, for reasons both greater and less than cosmic. Early that morning, looking out from the beach over the Mazacuni, as we called the union of the two great rivers, there was wind, yet no wind, as the sun prepared to lift above the horizon. The great soft-walled jungle was clear and distinct. Every reed at the landing had its unbroken counterpart in the still surface. But at the apex of the waters, the smoke of all the battles in the world had gathered, and upon this the sun slowly concentrated his powers, until he tore apart the cloak of mist, turning the dark surface, first to oxidized, and then to shining quicksilver. Instantaneously the same shaft of light touched the tips of the highest trees, and as if in response to a poised bâton, there broke forth that wonder of the world—the Zoroastrian chorus of tens of thousands of jungle creatures.

Over the quicksilver surface little individual

breezes wandered here and there. I could clearly see the beginning and the end of them, and one that drifted ashore and passed me felt like the lightest touch of a breath. One saw only the ripple on the water; one thought of invisible wings and trailing unseen robes.

With the increasing warmth the water-mist rose slowly, like a last quiet breath of night; and as it ascended,—the edges changing from silvery gray to grayish white,—it gathered close its shredded margins, grew smaller as it rose higher, and finally became a cloud. I watched it and wondered about its fate. Before the day was past, it might darken in its might, hurl forth thunders and jagged light, and lose its very substance in down-poured liquid. Or, after drifting idly high in air, the still-born cloud might garb itself in rich purple and gold for the pageant of the west, and again descend to brood over the coming marvel of another sunrise.

The tallest of bamboos lean over our low, lazy spread of bungalow; and late this very night, in the full moonlight, I leave my cot and walk down to the beach over a shadow carpet of Japanese filigree. The air over the white sand is as quiet and feelingless to my skin as complete, comfortable clothing. On one side is the dark river; on the other, the darker jungle full of gentle rustlings, low, velvety breaths of sound; and I slip into the water and swim out, out, out. Then I turn over and float along with the almost tangible moonlight flooding down on face and water. Suddenly the whole air is broken by the chorus of big red baboons, which rolls and tumbles toward me in masses of sound along the surface and goes trembling, echoing on over shore and jungle, till hurled back by the answering chorus of another clan. It stirs one to the marrow, for there is far more in it than the mere roaring of monkeys; and I turn uneasily, and slowly surge back toward the sand, overhand now, making companionable splashes.

And then again I stop, treading water softly, with face alone between river and sky; for the monkeys have ceased, and very faint and low, but blended in wonderful minor harmony, comes another chorus—from three miles down the river: the convicts singing hymns in their cells at midnight. And I ground gently and sit in the silvered shadows with little bewildered shrimps flicking against me, and unlanguaged thoughts come and go—impossible similes, too poignant phrases to be stopped and fettered with words, and I am neither scientist nor man nor naked organism, but just mind. With the coming of silence I look around and again consciously take in the scene. I am very glad to be alive, and to know that the possible dangers of jungle and water have not kept me armed and indoors. I feel, somehow, as if my very daring and gentle slipping-off of all signs of dominance and protection on entering into this realm had made friends of all the rare but possible serpents and scorpions, sting-rays and perai, vampires and electric eels. For a while I know the happiness of Mowgli.

And I think of people who would live more joyful lives in dense communities, who would be more tolerant, and more certain of straightforward friendship, if they could have as a background a fundamental hour of living such as this, a leaven for the rest of what, in comparison, seems mere existence.

At last I go back between the bamboos and their shadows, from unreal reality into a definiteness of cot and pajamas and electric torch. But wild nature still keeps touch with me; for as I write these lines, curled up on the edge of the cot, two vampires hawk back and forth so close that the wind from their wings dries my ink. And the soundness of my sleep is such that time does not exist between their last crepuscular squeak and the first wiry twittering of a blue tanager, in full sunshine, from a palm overhanging my beach.

V

A BIT OF USELESSNESS

A MOST admirable servant of mine once risked his life to reach a magnificent Bornean orchid, and tried to poison me an hour later when he thought I was going to take the plant away from him. This does not mean necessarily that we should look with suspicion upon all gardeners and lovers of flowers. It emphasizes, rather, the fact of the universal and deep-rooted appreciation of the glories of the vegetable kingdom. Long before the fatal harvest time, I am certain that Eve must have plucked a spray of apple blossoms with perfect impunity.

A vast amount of bad poetry and a much less quantity of excellent verse has been written about flowers, much of which follows to the letter Mark Twain's injunction about Truth. It must be admitted that the relations existing between the honeysuckle and the bee are basely practical and wholly selfish. A butterfly's admiration of a flower is no whit less than the blossom's conscious appreciation of its own beauties. There are ants which spend most of their life making gardens, knowing the uses of fertilizers, mulching, planting seeds, exercising patience, recognizing the time of ripeness, and gathering the edible fruit. But this is underground, and the ants are blind.

There is a bird, however—the bower bird of Australia—which appears to take real delight in bright things, especially pebbles and flowers for their own sake. Its little lean-to, or bower of sticks, which has been built in our own Zoological Park in New York City, is fronted by a cleared space, which is usually mossy. To this it brings its colorful treasures, sometimes a score of bright star blossoms, which are renewed when faded and replaced by others. All this has, probably, something to do with courtship, which should inspire a sonnet.

From the first pre-Egyptian who crudely scratched a lotus on his dish of clay, down to the jolly Feckenham men, the human race has given to flowers something more than idle curiosity, something less than mere earnest of fruit or berry.

At twelve thousand feet I have seen one of my Tibetans with nothing but a few shreds of straw between his bare feet and the snow, probe around the south edge of melting drifts until he found brilliant little primroses to stick behind his ears. I have been ushered into the little-used, musty best-parlor of a New England farmhouse, and seen fresh vases of homely, old-fashioned flowers—so recently placed for my edification, that drops of water still glistened like dewdrops on the dusty plush mat beneath. I have sat in the seat of honor of a Dyak communal house, looked up at the circle of all too recent heads, and seen a gay flower in each hollow eye socket, placed there for my approval. With a cluster of colored petals swaying in the breeze, one may at times bridge centuries or span the earth.

And now as I sit writing these words in my jungle laboratory, a small dusky hand steals around an aquarium and deposits a beautiful spray of orchids on my table. The little face appears, and I can distinguish the high cheek bones of Indian blood, the flattened nose and slight kink of negro, and the faint trace of white—probably of some long forgotten Dutch sailor, who came and went to Guiana, while New York City was still a browsing ground for moose.

So neither race nor age nor mélange of blood

can eradicate the love of flowers. It would be a wonderful thing to know about the first garden that ever was, and I wish that "Best Beloved" had demanded this. I am sure it was long before the day of dog, or cow, or horse, or even she who walked alone. The only way we can imagine it, is to go to some wild part of the earth, where are fortunate people who have never heard of seed catalogs or lawn mowers.

Here in British Guiana I can run the whole gamut of gardens, within a few miles of where I am writing. A mile above my laboratory up-river, is the thatched *benab* of an Akawai Indian —whose house is a roof, whose rooms are hammocks, whose estate is the jungle. Degas can speak English, and knows the use of my 28-gauge double barrel well enough to bring us a constant supply of delicious bushmeat—peccary, deer, monkey, bush turkeys and agoutis. But Grandmother has no language but her native Akawai. She is a good friend of mine, and we hold long conversations, neither of us bothering with the letter, but only the spirit of communication. She is a tiny person, bowed and wrinkled as only an old Indian squaw can be, always jolly and chuckling to herself, although

Degas tells me that the world is gradually darkening for her. And she vainly begs me to clear the film which is slowly closing over her eyes. She labors in a true landscape garden—the small circle wrested with cutlass and fire from the great jungle, and kept free only by constant cutting of the vines and lianas which creep out almost in a night, like sinister octopus tentacles, to strangle the strange upstarts and rejungle the bit of sunlit glade.

Although to the eye a mass of tangled vegetation, an Indian's garden may be resolved into several phases—all utterly practical, with color and flowers as mere by-products. First come the provisions, for if Degas were not hunting for me, and eating my rations, he would be out with bow and blowpipe, or fish-hooks, while the women worked all day in the cassava field. It is his part to clear and burn the forest, it is hers to grub up the rich mold, to plant and to weed. Plots and beds are unknown, for in every direction are fallen trees, too large to burn or be chopped up, and great sprawling roots. Between these, sprouts of cassava and banana are stuck, and the yams and melons which form the food of these primitive people. Cassava is as vital to these

Indians as the air they breathe. It is their wheat and corn and rice, their soup and salad and dessert, their ice and their wine, for besides being their staple food, it provides *casareep* which preserves their meat, and *piwarie* which, like excellent wine, brightens life for them occasionally, or dims it if overindulged in—which is equally true of food, or companionship, or the oxygen in the air we breathe.

Besides this cultivation, Grandmother has a small group of plants which are only indirectly concerned with food. One is *kunami,* whose leaves are pounded into pulp, and used for poisoning the water of jungle streams, with the surprising result that the fish all leap out on the bank and can be gathered as one picks up nuts. When I first visited Grandmother's garden, she had a few pitiful little cotton plants from whose stunted bolls she extracted every fiber and made a most excellent thread. In fact, when she made some bead aprons for me, she rejected my spool of cotton and chose her own, twisted between thumb and finger. I sent for seed of the big Sea Island cotton, and her face almost unwrinkled with delight when she saw the packets with seed larger than she had ever known.

Far off in one corner I make certain I have found beauty for beauty's sake, a group of exquisite caladiums and amaryllis, beautiful flowers and rich green leaves with spots and slashes of white and crimson. But this is the hunter's garden, and Grandmother has no part in it, perhaps is not even allowed to approach it. It is the *beena* garden—the charms for good luck in hunting. The similarity of the leaves to the head or other parts of deer or peccary or red-gilled fish, decides the most favorable choice, and the acrid, smarting juice of the tuber rubbed into the skin, or the hooks and arrows anointed, is considered sufficient to produce the desired result. Long ago I discovered that this demand for immediate physical sensation was a necessary corollary of doctoring, so I always give two medicines —one for its curative properties, and the other, bitter, sour, acid or anything disagreeable, for arousing and sustaining faith in my ability.

The Indian's medicine plants, like his true name, he keeps to himself, and although I feel certain that Grandmother had somewhere a toothache bush, or pain leaves—yarbs and simples for various miseries—I could never discover them. Half a dozen tall tobacco plants brought from the far interior, eked out the occasional tins of cigarettes in which Degas indulged, and always the flame-colored little buck-peppers lightened up the shadows of the *benab,* as hot to the palate as their color to the eye.

One day just as I was leaving, Grandmother led me to a palm nearby, and to one of its ancient frond-sheaths was fastened a small brown branch to which a few blue-green leaves were attached. I had never seen anything like it. She mumbled and touched it with her shriveled, bent fingers. I could understand nothing, and sent for Degas, who came and explained grudgingly, "Me no know what for—*toko-nook* just name—have got smell when yellow." And so at last I found the bit of uselessness, which, carried onward and developed in ages to come, as it had been elsewhere in ages past, was to evolve into botany, and back-yard gardens, and greenhouses, and wars of roses, and beautiful paintings, and music with a soul of its own, and verse more than human. To Degas the *toko-nook* was "just name," "and it was nothing more." But he was forgiven, for he had all unwittingly sowed the seeds of religion, through faith in his glowing caladiums. But Grandmother, though all the sunlight seemed dusk, and the dawn but as night, yet clung to her little plant, whose glory was that it was of no use whatsoever, but in months to come would be yellow, and would smell.

Farther down river, in the small hamlets of the boirianders—the people of mixed blood—the practical was still necessity, but almost every thatched and wattled hut had its swinging orchid branch, and perhaps a hideous painted tub with picketed rim, in which grew a golden splash of croton. This ostentatious floweritis might furnish a theme for a wholly new phase of the subject—for in almost every respect these people are less worthy human beings—physically, mentally and morally—than the Indians. But one cannot shift literary overalls for philosophical paragraphs in mid-article, so let us take the little river steamer down stream for forty miles to the coast of British Guiana, and there see what Nature herself does in the way of gardens. We drive twenty miles or more before we reach Georgetown, and the sides of the road are lined for most of the distance with huts and hovels of East Indian coolies and native Guiana negroes. Some are made of boxes, others of bark, more of thatch or rough-hewn boards and barrel staves, and some of split bamboo. But they resemble one another in several respects—all are ramshackle, all lean with the grace of Pisa, all have shutters and doors, so that at night they may be hermetically closed, and all are half-hidden in the folds of a curtain of flowers. The most shiftless, unlovely hovel, poised ready to return to its original chemical elements, is embowered in a mosaic of color, which in a northern garden would be worth a king's ransom—or to be strictly modern, should I not say a labor foreman's or a comrade's ransom!

The deep trench which extends along the front of these sad dwellings is sometimes blue with water hyacinths; next the water disappears beneath a maze of tall stalks, topped with a pink mist of lotus; then come floating lilies and more hyacinths. Wherever there is sufficient clear water, the wonderful curve of a cocoanut palm is etched upon it, reflection meeting palm, to form a dendritic pattern unequaled in human devising.

Over a hut of rusty oil-cans, bougainvillia stretches its glowing branches, sometimes cerise, sometimes purple, or allamanders fill the air with a golden haze from their glowing search-lights, either hiding the huts altogether, or softening their details into picturesque ruins. I remember one coolie dwelling which was dirtier and less habitable than the meanest stable, and all around it were hundreds upon hundreds of frangipanni blooms—the white and gold temple flowers of the East—giving forth of scent and color all that a flower is capable, to alleviate the miserable blot of human construction. Now and then a flamboyant tree comes into view, and as, at night, the head-lights of an approaching car eclipse all else, so this tree of burning scarlet draws eye and mind from adjacent human-made squalor. In all the tropics of the world I scarcely remember to have seen more magnificent color than in these unattended, wilful-grown gardens.

In tropical cities such as Georgetown, there are very beautiful private gardens, and the public one is second only to that of Java. But for the most part one is as conscious of the very dreadful borders of brick, or bottles, or conchs, as of the flowers themselves. Some one who is a master gardener will some day write of the possibilities of a tropical garden, which will hold the reader as does desire to behold the gardens of Carcassonne itself.

VI

GUINEVERE THE MYSTERIOUS

Again the Guiana jungle comes wonderfully to the eye and mysteriously to the mind; again my khakis and sneakers are skin-comfortable; again I am squatted on a pleasant mat of leaves in a miniature gorge, miles back of my Kartabo bungalow. Life elsewhere has already become unthinkable. I recall a place boiling with worried people, rent with unpleasing sounds, and beset with unsatisfactory pleasures. In less than a year I shall long for a sight of these worried people, my ears will strain to catch the unpleasing sounds, and I shall plunge with joy into the unsatisfactory pleasures. To-day, however, all these have passed from mind, and I settle down another notch, head snuggled on knees, and sway, elephant-fashion, with sheer joy, as a musky, exciting odor comes drifting, apparently by its own volition, down through the windless little gorge.

If I permit a concrete, scientific reaction, I

must acknowledge the source to be a passing bug,—a giant bug,—related distantly to our malodorous northern squash-bug, but emitting a scent as different as orchids' breath from grocery garlic. But I accept this delicate volatility as simply another pastel-soft sense-impression—as an earnest of the worthy, smelly things of old jungles. There is no breeze, no slightest shift of air-particles; yet down the gorge comes this cloud,—a cloud unsensible except to nostrils,—eddying as if swirling around the edges of leaves, riding on the air as gently as the low, distant crooning of great, sleepy jungle doves.

With two senses so perfectly occupied, sight becomes superfluous and I close my eyes. And straightway the scent and the murmur usurp my whole mind with a vivid memory. I am still squatting, but in a dark, fragrant room; and the murmur is still of doves; but the room is in the cool, still heart of the Queen's Golden Monastery in northern Burma, within storm-sound of Tibet, and the doves are perched among the glitter and tinkling bells of the pagoda roofs. I am squatting very quietly, for I am tired, after photographing carved peacocks and junglefowl in the marvelous fretwork of the outer balconies,

There are idols all about me—or so it would appear to a missionary; for my part, I can think only of the wonderful face of the old Lama who sits near me, a face peaceful with the something for which most of us would desert what we are doing, if by that we could attain it. Near him are two young priests, sitting as motionless as the Buddha in front of them.

After a half-hour of the strange thing that we call time, the Lama speaks, very low and very softly:

"The surface of the mirror is clouded with a breath."

Out of a long silence one of the neophytes replies, "The mirror can be wiped clear."

Again the world becomes incense and doves,—in the silence and peace of that monastery, it may have been a few minutes or a decade,—and the second Tibetan whispers, "There is no need to wipe the mirror."

When I have left behind the world of inharmonious colors, of polluted waters, of soot-stained walls and smoke-tinged air, the green of jungle comes like a cooling bath of delicate tints and shades. I think of all the green things I have loved—of malachite in matrix and table-top; of jade, not factory-hewn baubles, but age-mellowed signets, fashioned by lovers of their craft, and seasoned by the toying yellow fingers of generations of forgotten Chinese emperors—jade, as Dunsany would say, of the exact shade of the right color. I think too, of dainty emerald scarves that are seen and lost in a flash at a dance; of the air-cooled, living green of curling breakers; of a lonely light that gleams to starboard of an unknown passing vessel, and of the transparent green of northern lights that flicker and play on winter nights high over the garish glare of Broadway.

Now, in late afternoon, when I opened my eyes in the little gorge, the soft green vibrations merged insensibly with the longer waves of the doves' voices and with the dying odor. Soon the green alone was dominant; and when I had finished thinking of pleasant, far-off green things, the wonderful emerald of my great tree-frog of last year came to mind,—Gawain the mysterious,—and I wondered if I should ever solve his life.

In front of me was a little jungle rainpool. At the base of the miniature precipice of the gorge, this pool was a thing of clay. It was milky in consistence, from the roiling of suspended clay; and when the surface caught a glint of light and reflected it, only the clay and mud walls about came to the eye. It was a very regular pool, a man's height in diameter, and, for all I knew, from two inches to two miles deep. I became absorbed in a sort of subaquatic mirage, in which I seemed to distinguish reflections beneath the surface. My eyes refocused with a jerk, and I realized that something had unconsciously been perceived by my rods and cones, and short-circuited to my duller brain. Where a moment before was an unbroken translucent surface, were now thirteen strange beings who had appeared from the depths, and were mumbling oxygen with trembling lips.

In days to come, through all the months, I should again and again be surprised and cheated and puzzled—all phases of delight in the beings who share the earth's life with me. This was one of the first of the year, and I stiffened into one large eye.

I did not know whether they were fish, fairy shrimps, or frogs; I had never seen anything like them, and they were wholly unexpected. I so much desired to know what they were, that I sat quietly—as I enjoy keeping a treasured letter to the last, or reserving the frosting until the cake is eaten. It occurred to me that, had it not been for the Kaiser, I might have been forbidden this mystery; a chain of occurrences: Kaiser—war—submarines—glass-shortage for dreadnoughts—mica port-holes needed—Guiana prospector—abandoned pits—rainy season—mysterious tenants—me!

When I squatted by the side of the pool, no sign of life was visible. Far up through the green foliage of the jungle I could see a solid ceiling of cloud, while beneath me the liquid clay of the pool was equally opaque and lifeless. As a seer watches the surface of his crystal ball, so I gazed at my six-foot circle of milky water. My shift forward was like the fall of a tree: it brought into existence about it a temporary circle of silence and fear—a circle whose periphery began at once to contract; and after a few minutes the gorge again accepted me as a part of its harmless self. A huge bee zoomed past, and just behind my head a hummingbird beat the air into a froth of sound, as vibrant as the richest tones of a cello. My concentrated interest seemed to become known to the life of the surrounding glade, and I was bombarded with sight, sound, and odor, as if on purpose to distract my attention. But I remained unmoved, and indications of rare and desirable beings passed unheeded.

A flotilla of little water-striders came rowing themselves along, racing for a struggling ant which had fallen into the milky quicksand. These were in my line of vision, so I watched them for a while, letting the corner of my eye keep guard for the real aristocrats of the milky sea—whoever they were. My eye was close enough, my elevation sufficiently low to become one with the water-striders, and to become excited over the adventures of these little petrels; and in my absorption I almost forgot my chief quest. As soaring birds seem at times to rest against the very substance of cloud, as if upheld by some thin lift of air, so these insects glided as easily and skimmed as swiftly upon the surface film of water. I did not know even the genus of this tropical form; but insect taxonomists have been particularly happy in their given names—I recalled *Hydrobates, Aquarius,* and *remigis.*

The spur-winged jacanas are very skilful in their dainty treading of water-lily leaves; but here were good-sized insects rowing about on the water itself. They supported themselves on the four hinder legs, rowing with the middle pair, and steering with the hinder ones, while the front limbs were held aloft ready for the seizing of prey. I watched three of them approach the ant, which was struggling to reach the shore, and the first to reach it hesitated not a moment, but leaped into the air from a take-off of mere aqueous surface film, landed full upon the drowning unfortunate, grasped it, and at the same instant gave a mighty sweep with its oars, to escape from its pursuing, envious companions. Off went the twelve dimples, marking the aquatic footprints of the trio of striders; and as the bearer of the ant dodged one of its own kind, it was suddenly threatened by a small, jet submarine of a diving beetle. At the very moment when the pursuit was hottest, and it seemed anybody's ant, I looked aside, and the little water-bugs passed from my sight forever—for scattered over the surface were seven strange, mumbling mouths. Close as I was, their nature still eluded me. At my slightest movement all vanished, not with the virile splash of a fish or the healthy roll and dip of a porpoise, but with a weird, vertical withdrawing—the seven dissolving into the milk to join their six fellows.

This was sufficient to banish further meditative surmising, and I crept swiftly to a point of vantage, and with sweep-net awaited their reappearance. It was five minutes before faint, discolored spots indicated their rising, and at least two minutes more before they actually disturbed the surface. With eight or nine in view, I dipped quickly and got nothing. Then I sank my net deeply and waited again. This time ten minutes passed, and then I swept deep and swiftly, and drew up the net with four flopping, struggling super-tadpoles. They struggled for only a moment, and then lay quietly waiting for what might be sent by the guardian of the fate of tadpoles—surely some quaint little god-relation of Neptune, Pan, and St. Vitus. Gently shunted into a glass jar, these surprising tads accepted the new environment with quiet philosophy; and when I reached the laboratory and transferred them again, they dignifiedly righted themselves in the swirling current, and hung in mid-aquarium, waiting—forever waiting.

It was difficult to think of them as tadpoles, when the word brought to mind hosts of little black wrigglers filling puddles and swamps of our northern country. These were slow-moving, graceful creatures, partly transparent, partly reflecting every hue of the spectrum, with broad, waving scarlet and hyaline fins, and strange, fish-like mouths and eyes. Their habits were as unpollywoglike as their appearance. I visited their micaceous pool again and again; and if I could have spent days instead of hours with them, no moment of ennui would have intervened.

My acquaintanceship with tadpoles in the past had not aroused me to enthusiasm in the matter of their mental ability; as, for example, the inmates of the next aquarium to that of the Redfins, where I kept a herd or brood or school of Short-tailed Blacks—pollywogs of the Giant Toad (*Bufo marinus*). At earliest dawn they swam aimlessly about and mumbled; at high noon they mumbled and still swam; at midnight they refused to be otherwise occupied. It was possible to alarm them; but even while they fled they mumbled.

In bodily form my Redfins were fish, but mentally they had advanced a little beyond the usual tadpole train of reactions, reaching forward toward the varied activities of the future amphibian. One noticeable thing was their segregation, whether in the mica pools, or in two other smaller ones near by, in which I found them. Each held a pure culture of Redfins, and I found that this was no accident, but aided and enforced by the tads themselves. Twice, while I watched them, I saw definite pursuit of an alien pollywog,—the larva of the Scarlet-thighed Leaf-walker (*Phyllobates inguinalis*),—which fled headlong. The second time the attack was so persistent that the lesser tadpole leaped from the water, wriggled its way to a damp heap of leaves, and slipped down between them. For tadpoles to take such action as this was as reasonable as for an orchid to push a fellow blossom aside on the approach of a fertilizing hawk-moth. This momentary co-operation, and the concerted elimination of the undesired tadpole, affected me as the thought of the first consciousness of power of synchronous rhythm coming to ape men: it seemed a spark of tadpole genius—an adumbration of possibilities which now would end in the dull consciousness of the future frog, but which might, in past ages, have been a vital link in the development of an ancestral Ereops.

My Redfins were assuredly no common tadpoles, and an intolerant pollywog offers worthy research for the naturalist. Straining their medium of its opacity, I drew off the clayey liquid and replaced it with the clearer brown, wallaba-stained water of the Mazaruni; and thereafter all their doings, all their intimacies, were at my mercy. I felt as must have felt the first aviator who flew unheralded over an oriental city, with its patios and house-roofs spread naked beneath him.

It was on one of the early days of observation that an astounding thought came to me—before I had lost perspective in intensive watching, before familiarity had assuaged some of the marvel of these super-tadpoles. Most of those in my jar were of a like size, just short of an inch; but one was much larger, and correspondingly gorgeous in color and graceful in movement. As she swept slowly past my line of vision, she turned and looked, first at me, then up at the limits of her world, with a slow deliberateness and a hint of expression which struck deep into my memory. Green came to mind,—something clad in a smock of emerald, with a waistcoat of mother-of-pearl, and great sprawling arms,—and I found myself thinking of Gawain, our mystery frog of a year ago, who came without warning, and withheld all the secrets of his life. And I glanced again at this super-tad,—as unlike her ultimate development as the grub is unlike the beetle,—and one of us exclaimed, "It is the same, or nearly, but more delicate, more beautiful; it must be Guinevere." And so, probably for the first time in the world, there came to be a pet tadpole, one with an absurd name which will forever be more significant to us than the term applied by a forgotten herpetologist many years ago.

And Guinevere became known to all who had to do with the laboratory. Her health and daily development and color-change were things to be inquired after and discussed; one of us watched her closely and made notes of her life, one painted every radical development of color and pattern, another photographed her, and another brought her delectable scum. She was waited upon as sedulously as a termite queen. And she rewarded us by living, which was all we asked.

It is difficult for a diver to express his emotions on paper, and verbal arguments with a dentist are usually one-sided. So must the spirit of a tadpole suffer greatly from handicaps of the flesh. A mumbling mouth and an uncontrollable, flagellating tail, connected by a pinwheel of intestine, are scant material wherewith to attempt new experiments, whereon to nourish aspirations. Yet the Redfins, as typified by Guinevere, have done both, and given time enough, they may emulate or surpass the achievements of larval axolotls, or the astounding egg-producing maggots of certain gnats, thus realizing all the possibilities of froghood while yet cribbed within the lowly casing of a pollywog.

In the first place Guinevere had ceased being positively thigmotactic, and, writing as a technical herpetologist, I need add no more. In fact, all my readers, whether Batrachologists or Casuals, will agree that this is an unheard-of achievement. But before I loosen the technical etymology and become casually more explicit, let me hold this term in suspense a moment, as I once did, fascinated by the sheer sound of the syllables, as they first came to my ears years ago in a university lecture. There is that of possibility in being positively thigmotactic which makes one dread the necessity of exposing and limiting its meaning, of digging down to its mathematically accurate roots. It could never be called a flower of speech: it is an over-ripe fruit rather: heavy-stoned, thin-fleshed—an essentially practical term. It is eminently suited to its purpose, and so widely used that my friend the editor must accept it; not looking askance as he did at my definition of a vampire as a vespertilial anæsthetist, or breaking into open but wholly ineffectual rebellion, at the past tense of the verb to candelabra. I admit that the conjugation

I candelabra
You candelabra
He candelabras

arouses a ripple of confusion in the mind; but it is far more important to use words than to parse them, anyway, so I acclaim perfect clarity for "The fireflies candelabraed the trees!"

Not to know the precise meaning of being positively thigmotactic is a stimulant to the imagination, which opens the way to an entire essay on the disadvantages of education—a thought once strongly aroused by the glorious red-and-gold hieroglyphic signs of the Peking merchants—signs which have always thrilled me more than the utmost efforts of our modern psychological advertisers.

Having crossed unconsciously by such a slender etymological bridge from my jungle tadpole to China, it occurs to me that the Chinese are the most positively thigmotactic people in the world. I have walked through block after block of subterranean catacombs, beneath city streets which were literally packed full of humanity, and I have seen hot mud pondlets along the Min River wholly eclipsed by shivering Chinamen packed sardinewise, twenty or thirty in layers, or radiating like the spokes of a great wheel which has fallen into the mud.

From my brood of Short-tailed Blacks, a half-dozen tadpoles wandered off now and then, each scum-mumbling by himself. Shortly his positivism asserted itself and back he wriggled, twisting in and out of the mass of his fellows, or at the approach of danger nuzzling into the dead leaves at the bottom, content only with the feeling of something pressing against his sides and tail. His physical make-up, simple as it is, has proved perfectly adapted to this touch system of life: flat-bottomed, with rather narrow, paddle-shaped tail-fins which, beginning well back of the body, interfere in no way with the pollywog's instincts, he can thigmotact to his heart's content. His eyes are also adapted to looking upward, discerning dimly dangers from above, and whatever else catches the attention of a bottom-loving pollywog. His mouth is well below, as best suits bottom mumbling.

Compared with these *polloi* pollywogs, Redfins were as hummingbirds to quail. Their very origin was unique; for while the toad tadpoles wriggled their way free from egg gelatine deposited in the water itself, the Redfins were literally rained down. Within a folded leaf the parents left the eggs—a leaf carefully chosen as overhanging a suitable ditch, or pit, or puddle. If all signs of weather and season failed and a sudden drought set in, sap would dry, leaf would shrivel, and the pitiful gamble for life of the little jungle frogs would be lost; the spoonful of froth would collapse bubble by bubble, and, finally, a thin dry film on the brown leaf would in turn vanish, and Guinevere and her companions would never have been.

But untold centuries of unconscious necessity have made these tree-frogs infallible weather prophets, and the liberating rain soon sifted through the jungle foliage. In the streaming drops which funneled from the curled leaf, tadpole after tadpole hurtled downward and splashed headlong into the water; their parents and the rain and gravitation had performed their part, and from now on fate lay with the super-tads themselves—except when a passing naturalist brought new complications, new demands of Karma, as strange and unpredictable as if from another planet or universe.

Only close examination showed that these were tadpoles, not fish, judged by the staring eyes, and broad fins stained above and below with orange-scarlet—colors doomed to oblivion in the native, milky waters, but glowing brilliantly in my aquarium. Although they were provided with such an expanse of fin, the only part used for ordinary progression was the extreme tip, a mere threadlike streamer, which whipped in never-ending spirals, lashing forward, backward, and sideways. So rapid was this motion, and so short the flagellum, that the tadpole did not even tremble or vibrate as it moved, but forged steadily onward, without a tremor.

The head was buffy yellow, changing to bittersweet orange back of the eyes and on the gills. The body was dotted with a host of minute specks of gold and silver. On the sides and below, this gave place to a rich bronze, and then to a clear, iridescent silvery blue. The eye proper was silvery white, but the upper part of the eyeball fairly glowed with color. In front it was jet black flecked with gold, merging behind into a brilliant blue. Yet this patch of jeweled tissue was visible only rarely as the tadpole turned forward, and in the opaque liquid of the mica pool must have ever been hidden. And even if plainly seen, of what use was a shred of rainbow to a sexless tadpole in the depths of a shady pool!

With high-arched fins, beginning at neck and throat, body compressed as in a racing yacht, there could be no bottom life for Guinevere. Whenever she touched a horizontal surface,—whether leaf or twig,—she careened; when she sculled through a narrow passage in the floating algæ, her fins bent and rippled as they were pressed bodywards. So she and her fellow brood lived in mid-aquarium, or at most rested lightly against stem or glass, suspended by gentle suction of the complex mouth. Once, when I inserted a long streamer of delicate water-weed, it remained upright, like some strange tree of carboniferous memory. After an hour I found this the perching-place of fourteen Redfin tads, and at the very summit was Guinevere. The rest were arranged nearly in altitudinal size—two large tadpoles being close below Guinevere, and a bevy of six tiny chaps lowest down. All were lightly poised, swaying in mid-water, at a gently sloping angle, like some unheard-of, orange-stained, aquatic autumn foliage.

For two weeks Guinevere remained almost as I have described her, gaining slightly in size, but with little alteration of color or pattern. Then came the time of the great change: we felt it to be imminent before any outward signs indicated its approach. And for four more days there was no hint except the sudden growth of the hind legs. From tiny dangling appendages with minute toes and indefinite knees, they enlarged and bent, and became miniature but perfect frog's limbs.

She had now reached a length of two inches, and her delicate colors and waving fins made her daily more marvelous. The strange thing about the hind limbs was that, although so large and perfect, they were quite useless. They could not even be unflexed; and other mere pollywogs near by were wriggling toes, calves, and thighs while yet these were but imperfect buds. When she dived suddenly, the toes occasionally moved a little; but as a whole, they merely sagged and drifted like some extraneous things entangled in the body.

Smoothly and gracefully Guinevere moved about the aquarium. Her gills lifted and closed rhythmically—twice as slowly as compared with the three or four times every second of her breathless young tadpolehood. Several times on the fourteenth day, she came quietly to the surface for a gulp of air.

Looking at her from above, two little bulges were visible on either side of the body—the ensheathed elbows pressing outward. Twice, when she lurched forward in alarm, I saw these front limbs jerk spasmodically; and when she was resting quietly, they rubbed and pushed impatiently against their mittened tissue.

And now began a restless shifting, a slow, strange dance in mid-water, wholly unlike any movement of her smaller companions; up and down, slowly revolving on oblique planes, with rhythmical turns and sinkings—this continued for an hour, when I was called for lunch. And as if to punish me for this material digression and desertion, when I returned, in half an hour, the miracle had happened.

Guinevere still danced in stately cadence, with the other Redfins at a distance going about their several businesses. She danced alone—a dance of change, of happenings of tremendous import, of symbolism as majestic as it was age-old. Here in this little glass aquarium the tadpole Guinevere had just freed her arms—she, with waving scarlet fins, watching me with lidless white and staring eyes, still with fish-like, fin-bound body. She danced upright, with new-born arms folded across her breast, tail-tip flagellating frenziedly, stretching long fingers with disks like cymbals, reaching out for the land she had never trod, limbs flexed for leaps she had never made.

A few days before and Guinevere had been a fish, then a helpless biped, and now suddenly, somewhere between my salad and coffee, she became an aquatic quadruped. Strangest of all, her hands were mobile, her feet useless; and when the dance was at an end, and she sank slowly to the bottom, she came to rest on the very tips of her two longest fingers; her legs and toes still drifting high and useless. Just before she ceased, her arms stretched out right froggily, her weird eyes rolled about, and she gulped a mighty gulp of the strange thin medium that covered the surface of her liquid home.

At midnight of this same day only three things existed in the world—on my table I turned from the *Bhagavad-Gita* to Drinkwater's *Reverie* and back again; then I looked up to the jar of clear water and watched Guinevere hovering motionless. At six the next morning she was crouched safely on a bit of paper a foot from the aquarium. She had missed the open window, the four-foot drop to the floor, and a neighboring aquarium stocked with voracious fish: surely the gods of pollywogs were kind to me. The great fins were gone—dissolved into blobs of dull pink; the tail was a mere stub, the feet drawn close, and a glance at her head showed that Guinevere had become a frog almost within an hour. Three things I hastened to observe: the pupils of her eyes were vertical, revealing her genus *Phyllomedusa* (making apt our choice of the feminine); by a gentle urging I saw that the first and second toes were equal in length; and a glance at her little humped back showed a scattering of white calcareous spots, giving the clue to her specific personality—*bicolor:* thus were we introduced to *Phyllomedusa bicolor,* alias Guinevere, and thus was established beyond doubt her close relationship to Gawain.

During that first day, within three hours, during most of which I watched her closely, Guinevere's change in color was beyond belief. For an hour she leaped from time to time; but after that, and for the rest of her life, she crept in strange unfroglike fashion, raised high on all four limbs, with her stubby tail curled upward, and reaching out one weird limb after another. If one's hand approached within a foot, she saw it and stretched forth appealing, skinny fingers.

At two o'clock she was clad in a general cinnamon buff; then a shade of glaucous green began to creep over head and upper eyelids, onward over her face, finally coloring body and limbs. Beneath, the little pollyfrog fairly glowed with bright apricot orange, throat and tail amparo purple, mouth green, and sides rich pale blue. To this maze of color we must add a strange, new expression, born of the prominent eyes, together with the line of the mouth extending straight back with a final jeering, upward lift; in front, the lower lip thick and protruding, which, with the slanting eyes, gave a leering, devilish smirk, while her set, stiff, exact posture compelled a vivid thought of the sphinx. Never have I seen such a remarkable combination. It fascinated us. We looked at Guinevere, and then at the tadpoles swimming quietly in their tank, and evolution in its wildest conceptions appeared a tame truism.

This was the acme of Guinevere's change, the pinnacle of her development. Thereafter her transformations were rhythmical, alternating with the day and night. Through the nights of activity she was garbed in rich, warm brown. With the coming of dawn, as she climbed slowly upward, her color shifted through chestnut to maroon; this maroon then died out on the mid-back to a delicate, dull violet-blue, which in turn became obscured in the sunlight by turquoise, which crept slowly along the sides. Carefully and laboriously she clambered up, up to the topmost frond, and there performed her little toilet, scraping head and face with her hands, passing the hinder limbs over her back to brush off every grain of sand. The eyes had meanwhile lost their black-flecked, golden, nocturnal iridescence, and had gradually paled to a clear silvery blue, while the great pupil of darkness narrowed to a slit.

Little by little her limbs and digits were drawn

in out of sight, and the tiny jeweled being crouched low, hoping for a day of comfortable clouds, a little moisture, and a swift passage of time to the next period of darkness, when it was fitting and right for Guineveres to seek their small meed of sustenance, to grow to frog's full estate, and to fulfil as well as might be what destiny the jungle offered. To unravel the meaning of it all is beyond even attempting. The breath of mist ever clouds the mirror, and only as regards a tiny segment of the life-history of Guinevere can I say, "There is no need to wipe the mirror."

VII

A JUNGLE LABOR-UNION

PTERODACTYL PUPS led me to the wonderful Attas—the most astounding of the jungle labor-unions. We were all sitting on the Mazaruni bank, the night before the full moon, immediately in front of my British Guiana laboratory. All the jungle was silent in the white light, with now and then the splash of a big river fish. On the end of the bench was the monosyllabic Scot, who ceased the exquisite painting of mora buttresses and jungle shadows only for the equal fascination of searching bats for parasites. Then the great physician, who had come six thousand miles to peer into the eyes of birds and lizards in my dark-room, working with a gentle hypnotic manner that made the little beings seem to enjoy the experience. On my right sat an army captain, who had given more thought to the possible secrets of French chaffinches than to the approaching barrage. There was also the artist, who could draw a lizard's head like a Japanese print, but preferred to depict impressionistic Laocoön roots.

These and others sat with me on the long bench and watched the moonpath. The conversation had begun with possible former life on the moon, then shifted to Conan Doyle's *The Lost World,* based on the great Roraima plateau, a hundred and fifty miles west of where we were sitting. Then we spoke of the amusing world-wide rumor, which had started no one knows how, that I had recently discovered a pterodactyl. One delightful result of this had been a letter from a little English girl, which would have made a worthy chapter-subject for *Dream Days.* For years she and her little sister had peopled a wood near her home with pterodactyls, but had somehow never quite seen one; and would I tell her a little about them—whether they had scales, or made nests; so that those in the wood might be a little easier to recognize.

When strange things are discussed for a long time, in the light of a tropical moon, at the edge of a dark, whispering jungle, the mind becomes singularly imaginative and receptive; and, as I looked through powerful binoculars at the great suspended globe, the dead craters and precipices became very vivid and near. Suddenly, without warning, there flapped into my field, a huge shapeless creature. It was no bird, and there was nothing of the bat in its flight—the wings moved with steady rhythmical beats, and drove it straight onward. The wings were skinny, the body large and of a pale ashy hue. For a moment I was shaken. One of the others had seen it, and he, too, did not speak, but concentrated every sense into the end of the little tubes. By the time I had begun to find words, I realized that a giant fruit bat had flown from utter darkness across my line of sight; and by close watching we soon saw others. But for a very few seconds these Pterodactyl Pups, as I nicknamed them, gave me all the thrill of a sudden glimpse into the life of past ages. The last time I had seen fruit bats was in the gardens of Perideniya, Ceylon. I had forgotten that they occurred in Guiana, and was wholly unprepared for the sight of bats a yard across, with a heron's flight, passing high over the Mazaruni in the moonlight.

The talk ended on the misfortune of the configuration of human anatomy, which makes sky-searching so uncomfortable a habit. This outlook was probably developed to a greater extent during the war than ever before; and I can remember many evenings in Paris and London when a sinister half-moon kept the faces of millions turned searchingly upward. But whether in city or jungle, sky-scanning is a neck-aching affair.

The following day my experience with the Pterodactyl Pups was not forgotten, and as a direct result of looking out for soaring vultures and eagles, with hopes of again seeing a white-plumaged King and the regal Harpy, I caught sight of a tiny mote high up in mid-sky. I thought at first it was a martin or swift; but it descended, slowly spiraling, and became too small for any bird. With a final, long, descending curve, it alighted in the compound of our bungalow laboratory and rested quietly—a great queen of the leaf-cutting Attas returning from her marriage flight. After a few minutes she stirred, walked a few steps, cleaned her antennæ, and searched nervously about on the sand. A foot away was a tiny sprig of indigo, the offspring of some seed planted two or three centuries ago by a thrifty Dutchman. In the shade of its three leaves the insect paused, and at once began scraping at the sand with her jaws. She loosened grain after grain, and as they came free they were moistened, agglutinated, and pressed back against her fore-legs. When at last a good-sized ball was formed, she picked it up, turned around and, after some fussy indecision, deposited it on the sand behind her. Then she returned to the very shallow, round depression, and began to gather a second ball.

I thought of the first handful of sand thrown out for the base of Cheops, of the first brick placed in position for the Great Wall, of a fresh-cut trunk, rough-hewn and squared for a log-cabin on Manhattan; of the first shovelful of earth flung out of the line of the Panama Canal. Yet none seemed worthy of comparison with even what little I knew of the significance of this ant's labor, for this was earnest of what would make trivial the engineering skill of Egyptians, of Chinese patience, of municipal pride and continental schism.

Imagine sawing off a barn-door at the top of a giant sequoia, growing at the bottom of the Grand Cañon, and then, with five or six children clinging to it, descending the tree, and carrying it up the cañon walls against a subway rush of rude people, who elbowed and pushed blindly against you. This is what hundreds of leaf-cutting ants accomplish daily, when cutting leaves from a tall bush, at the foot of the bank near the laboratory.

There are three dominant labor-unions in the jungle, all social insects, two of them ants, never interfering with each other's field of action, and all supremely illustrative of conditions resulting from absolute equality, free-and-equalness, communalism, socialism carried to the (forgive me!) anth power. The Army Ants are carnivorous, predatory, militant nomads; the Termites are vegetarian scavengers, sedentary, negative and provincial; the Attas, or leaf-cutting ants, are vegetarians, active and dominant, and in many ways the most interesting of all.

The casual observer becomes aware of them through their raids upon gardens; and indeed the Attas are a very serious menace to agriculture in many parts of the tropics, where their nests, although underground, may be as large as a house and contain millions of individuals. While their choice among wild plants is exceedingly varied, it seems that there are certain things they will not touch; but when any human-reared flower, vegetable, shrub, vine, or tree is planted, the Attas rejoice, and straightway desert the native vegetation to fall upon the newcomers. Their whims and irregular feeding habits make it difficult to guard against them. They will work all round a garden for weeks, perhaps pass through it *en route* to some tree that they are defoliating, and then suddenly, one night, every Atta in the world seems possessed with a desire to work havoc, and at daylight the next morning, the garden looks like winter stubble—a vast expanse of stems and twigs, without a single remaining leaf. Volumes have been written, and a whole chemist's shop of deadly concoctions devised, for combating these ants, and still they go steadily on, gathering leaves which, as we shall see, they do not even use for food.

Although essentially a tropical family, Attas have pushed as far north as New Jersey, where they make a tiny nest, a few inches across, and bring to it bits of pine needles.

In a jungle Baedeker, we should double-star these insects, and paragraph them as "*Atta,* named by Fabricius in 1804; the Kartabo species, *cephalotes;* Leaf-cutting or Cushie or Parasol Ants; very abundant. *Atta,* a subgenus of *Atta,* which is a genus of *Attii,* which is a tribe of *Myr-*

micinæ, which is a subfamily of *Formicidæ,*" etc.

With a feeling of slightly greater intimacy, of mental possession, we set out, armed with a name of one hundred and seventeen years' standing, and find a big Atta worker carving away at a bit of leaf, exactly as his ancestors had done for probably one hundred and seventeen thousand years.

We gently lift him from his labor, and a drop of chloroform banishes from his ganglia all memory of the hundred thousand years of pruning. Under the lens his strange personality becomes manifest, and we wonder whether the old Danish zoölogist had in mind the slender toe-tips which support him, or in a chuckling mood made him a namesake of C. Quintius Atta. A close-up shows a very comic little being, encased in a prickly, chestnut-colored armor, which should make him fearless in a den of a hundred anteaters. The front view of his head is a bit mephistophelian, for it is drawn upward into two horny spines; but the side view recalls a little girl with her hair brushed very tightly up and back from her face.

The connection between Atta and the world

about him is furnished by this same head: two huge, flail-shaped attennæ arching up like aerial, detached eyebrows—vehicles, through their golden pile, of senses which foil our most delicate tests. Outside of these are two little shoe-button eyes; and we are not certain whether they reflect to the head ganglion two or three hundred bits of leaf, or one large mosaic leaf. Below all is swung the pair of great scythes, so edged and hung that they can function as jaws, rip-saws, scissors, forceps, and clamps. The thorax, like the head of a titanothere, bears three pairs of horns—a great irregular expanse of tumbled, rock-like skin and thorn, a foundation for three pairs of long legs, and sheltering somewhere in its heart a thread of ant-life; finally, two little pedicels lead to a rounded abdomen, smaller than the head. This Third-of-an-inch is a worker Atta to the physical eye; and if we catch another, or ten, or ten million, we find that some are small, others much larger, but that all are cast in the same mold, all indistinguishable except, perhaps, to the shoe-button eyes.

When a worker has traveled along the Atta trails, and has followed the temporary mob-instinct and climbed bush or tree, the same irresistible force drives him out upon a leaf. Here, apparently, instinct slightly loosens its hold, and he seems to become individual for a moment, to look about, and to decide upon a suitable edge or corner of green leaf. But even in this he probably has no choice. At any rate, he secures a good hold and sinks his jaws into the tissue. Standing firmly on the leaf, he measures his distance by cutting across a segment of a circle, with one of his hind feet as a center. This gives a very true curve, and provides a leaf-load of suitable size. He does not scissor his way across, but bit by bit sinks the tip of one jaw, hook-like, into the surface, and brings the other up to it, slicing through the tissue with surprising ease. He stands upon the leaf, and I always expect to see him cut himself and his load free, Irishman-wise. But one or two of his feet have invariably secured a grip on the plant, sufficient to hold him safely. Even if one or two of his fellows are at work farther down the leaf, he has power enough in his slight grip to suspend all until they have finished and clambered up over him with their loads.

Holding his bit of leaf edge-wise, he bends his head down as far as possible, and secures a strong purchase along the very rim. Then, as he raises his head, the leaf rises with it, suspended high over his back, out of the way. Down the stem or tree-trunk he trudges, head first, fighting with gravitation, until he reaches the ground. After a few feet, or, measured by his stature, several hundred yards, his infallible instinct guides him around pebble boulders, mossy orchards, and grass jungles to a specially prepared path.

Thus in words, in sentences, we may describe the cutting of a single leaf; but only in the imagination can we visualize the cell-like or crystal-like duplication of this throughout all the great forests of Guiana and of South America. As I write, a million jaws snip through their stint; as you read, ten million Attas begin on new bits of leaf. And all in silence and in dim light, legions passing along the little jungle roads, unending lines of trembling banners, a political parade of ultra socialism, a procession of chlorophyll floats illustrating unreasoning unmorality, a fairy replica of "Birnam Forest come to Dunsinane."

In their leaf-cutting, Attas have mastered mass, but not form. I have never seen one cut off a piece too heavy to carry, but many a hard-sliced bit has had to be deserted because of the configuration of the upper edge. On almost any trail, an ant can be found with a two-inch stem of grass, attempting to pass under a twig an inch overhead. After five or ten minutes of pushing, backing, and pulling, he may accidentally march off to one side, or reach up and climb over; but usually he drops his burden. His little works have been wound up, and set at the mark "home"; and though he has now dropped the prize for which he walked a dozen ant-miles, yet any idea of cutting another stem, or of picking up a slice of leaf from those lying along the trail, never occurs to him. He sets off homeward, and if any emotion of sorrow, regret, disappointment, or secret relief troubles his ganglia, no trace of it appears in antennæ, carriage, or speed. I can very readily conceive of his trudging sturdily all the way back to the nest, entering it, and going to the place where he would have dumped his load, having fulfilled his duty in the spirit at least. Then, if there comes a click in his internal time-clock, he may set out upon another quest—more cabined, cribbed, and confined than any member of a Cook's tourist party.

I once watched an ant with a piece of leaf

which had a regular shepherd's crook at the top, and if his adventures of fifty feet could have been caught on a moving-picture film, Charlie Chaplin would have had an arthropod rival. It hooked on stems and pulled its bearer off his feet, it careened and ensnared the leaves of other ants, at one place mixing up with half a dozen. A big thistledown became tangled in it, and well-nigh blew away with leaf and all; hardly a foot of his path was smooth-going. But he persisted, and I watched him reach the nest, after two hours of tugging and falling and interference with traffic.

Occasionally an ant will slip in crossing a twiggy crevasse, and his leaf become tightly wedged. After sprawling on his back and vainly clawing at the air for a while, he gets up, brushes off his antennæ, and sets to work. For fifteen minutes I have watched an Atta in this predicament, stodgily endeavoring to lift his leaf while standing on it at the same time. The equation of push equaling pull is fourth dimensional to the Attas.

With all this terrible expenditure of energy, the activities of these ants are functional within very narrow limits. The blazing sun causes them to drop their burdens and flee for home; a heavy wind frustrates them, for they cannot reef. When a gale arises and sweeps an exposed portion of the trail, their only resource is to cut away all sail and heave it overboard. A sudden downpour reduces a thousand banners and waving, bright-colored petals to débris, to be trodden under foot. Sometimes, after a ten-minute storm, the trails will be carpeted with thousands of bits of green mosaic, which the outgoing hordes will trample in their search for more leaves. On a dark night little seems to be done; but at dawn and dusk, and in the moonlight or clear starlight, the greatest activity is manifest.

Attas are such unpalatable creatures that they are singularly free from dangers. There is a tacit armistice between them and the other labor-unions. The army ants occasionally make use of their trails when they are deserted; but when the two great races of ants meet, each antennæs the aura of the other, and turns respectfully aside. When termites wish to traverse an Atta trail, they burrow beneath it, or build a covered causeway across, through which they pass and repass at will, and over which the Attas trudge, uncaring and unconscious of its significance.

Only creatures with the toughest of digestions would dare to include these prickly, strong-jawed, meatless insects in a bill of fare. Now and then I have found an ani, or black cuckoo, with a few in its stomach: but an ani can swallow a stinging-haired caterpillar and enjoy it. The most consistent feeder upon Attas is the giant marine toad. Two hundred Attas in a night is not an uncommon meal, the exact number being verifiable by a count of the undigested remains of heads and abdomens. *Bufo marinus* is the gardener's best friend in this tropic land, and besides, he is a gentleman and a philosopher, if ever an amphibian was one.

While the cutting of living foliage is the chief aim in life of these ants, yet they take advantage of the flotsam and jetsam along the shore, and each low tide finds a column from some nearby nest salvaging flowerets, leaves, and even tiny berries. A sudden wash of tide lifts a hundred ants with their burdens and then sets them down again, when they start off as if nothing had happened.

The paths or trails of the Attas represent very remarkable feats of engineering, and wind about through jungle and glade for surprising distances. I once traced a very old and wide trail for well over two hundred yards. Taking little Third-of-an-inch for a type (although he would rank as a rather large Atta), and comparing him with a six-foot man, we reckon this trail, ant-ratio, as a full twenty-five miles. Belt records a leaf-cutter's trail half a mile long, which would mean that every ant that went out, cut his tiny bit of leaf, and returned, would traverse a distance of a hundred and sixteen miles. This was an extreme; but our Atta may take it for granted, speaking antly, that once on the home trail, he has, at the least, four or five miles ahead of him.

The Atta roads are clean swept, as straight as possible, and very conspicuous in the jungle. The chief high-roads leading from very large nests are a good foot across, and the white sand of their beds is visible a long distance away. I once knew a family of opossums living in a stump in the center of a dense thicket. When they left at evening, they always climbed along as far as an Atta trail, dropped down to it, and followed it for twenty or thirty yards. During the rains I have occasionally found tracks of agoutis and deer in these roads. So it would be very possible for the Attas to lay the foundation for an animal trail, and this, *à la* calf-path, for the street of a future city.

The part that scent plays in the trails is evidenced if we scatter an inch or two of fresh sand across the road. A mass of ants banks against the strange obstruction on both sides, on the one hand a solid phalanx of waving green banners, and on the other a mob of empty-jawed workers with wildly waving antennæ. Scouts from both sides slowly wander forward, and finally reach one another and pass across. But not for ten minutes does anything like regular traffic begin again.

When carrying a large piece of leaf, and traveling at a fair rate of speed, the ants average about a foot in ten seconds, although many go the same distance in five. I tested the speed of an Atta, and then I saw that its leaf seemed to have a peculiar-shaped bug upon it, and picked it up with its bearer. Finding the blemish to be only a bit of fungus, I replaced it. Half an hour later I was seated by a trail far away, when suddenly my ant with the blemished spot appeared. It was unmistakable, for I had noticed that the spot was exactly that of the Egyptian symbol of life. I paced the trail, and found that seventy yards away it joined the spot where I had first seen my friend. So, with occasional spurts, he had done two hundred and ten feet in thirty minutes, and this in spite of the fact that he had picked up a supercargo.

Two parts of hydrogen and one of oxygen, under the proper stimulus, invariably result in water; two and two, considered calmly and without passion, combine into four; the workings of instinct, especially in social insects, is so mechanical that its results can almost be demonstrated in formula; and yet here was my Atta leaf-carrier burdened with a minim. The worker Attas vary greatly in size, as a glance at a populous trail will show. They have been christened *macrergates, desmergates* and *micrergates;* or we may call the largest Maxims, the average middle class Mediums, and the tiny chaps Minims, and all have more or less separate functions in the ecology of the colony. The Minims are replicas in miniature of the big chaps, except that their armor is pale cinnamon rather than chestnut. Although they can bite ferociously, they are too small to cut through leaves, and they have very definite duties in the nest; yet they are found with every leaf-cutting gang, hastening along with their larger brethren, but never doing anything, that I could detect, at their journey's end. I have a suspicion that the little Minims, who are very numerous, function as light cavalry; for in case of danger they are as eager at attack as the great soldiers, and the leaf-cutters, absorbed in their arduous labor, would benefit greatly from the immunity ensured by a flying corps of their little bulldog comrades.

I can readily imagine that these nestling Minims become weary and foot-sore (like bank-clerks guarding a reservoir), and if instinct allows such abominable individuality, they must often wish themselves back at the nest, for every mile of a Medium is three miles to them.

Here is where our mechanical formula breaks down; for, often, as many as one in every five leaves that pass bears aloft a Minim or two, clinging desperately to the waving leaf and getting a free ride at the expense of the already overburdened Medium. Ten is the extreme number seen, but six to eight Minims collected on a single leaf is not uncommon. Several times I have seen one of these little banner-riders shift deftly from leaf to leaf, when a swifter carrier passed by, as a circus bareback rider changes steeds at full gallop.

Once I saw enacted above ground, and in the light of day, something which may have had its roots in an *anlage* of divine discontent. If I were describing the episode half a century ago, I should entitle it, "The Battle of the Giants, or Emotion Enthroned." A quadruple line of leaf-carriers was disappearing down a hole in front of the laboratory, bumped and pushed by an out-pouring, empty-jawed mass of workers. As I watched them, I became aware of an area of great excitement beyond the hole. Getting down as nearly as possible to ant height, I witnessed a terrible struggle. Two giants—of the largest soldier Maxim caste—were locked in each other's jaws, and to my horror, I saw that each had lost his abdomen. The antennæ and the abdomen petiole are the only vulnerable portions of an Atta, and long after he has lost these apparently dispensable portions of his anatomy, he is able to walk, fight, and continue an active but erratic life. These mighty-jawed fellows seem never to come to the surface unless danger threatens; and my mind went down into the black, musty depths, where it is the duty of these soldiers to walk about and wait for trouble. What could have raised the ire of such stolid neuters against one another? Was it sheer lack of something to do? or was there a cell or two of the winged caste lying fallow within their bodies, which, stirring at last, inspired a will to battle, a passing echo of romance, of the activities of the male Atta?

Their unnatural combat had stirred scores of smaller workers to the highest pitch of excitement. Now and then, out of the mêlée, a Medium would emerge, with a tiny Minim in his jaws. One of these carried his still living burden many feet away, along an unused trail, and dropped it. I examined the small ant, and found that it had lost an antenna, and its body was crushed. When the ball of fighters cleared, twelve small ants were seen clinging to the legs and heads of the mutilated giants, and now and then these would loosen their hold on each other, turn, and crush one of their small tormenters. Several times I saw a Medium rush up and tear a small ant away, apparently quite insane with excitement.

Occasionally the least exhausted giant would

stagger to his four and a half remaining legs, hoist his assailant, together with a mass of the midgets, high in air, and stagger for a few steps, before falling beneath the onrush of new attackers. It made me wish to help the great insect, who, for aught I knew, was doomed because he was different—because he had dared to be an individual.

I left them struggling there, and half an hour later, when I returned, the episode was just coming to a climax. My Atta hero was exerting his last strength, flinging off the pile that assaulted him, fighting all the easier because of the loss of his heavy body. He lurched forward, dragging the second giant, now dead, not toward the deserted trail or the world of jungle around him, but headlong into the lines of stupid leaf-carriers, scattering green leaves and flower-petals in all directions. Only when dozens of ants threw themselves upon him, many of them biting each other in their wild confusion, did he rear up for the last time, and, with the whole mob, rolled down into the yawning mouth of the Atta nesting-hole, disappearing from view, and carrying with him all those hurrying up the steep sides. It was a great battle. I was breathing fast with sympathy, and whatever his cause, I was on his side.

The next day both giants were lying on the old, disused trail; the revolt against absolute democracy was over; ten thousand ants passed to and fro without a dissenting thought, or any thought, and the Spirit of the Attas was content.

VIII

THE ATTAS AT HOME

Clambering through white, pasty mud which stuck to our boots by the pound, peering through bitter cold mist which seemed but a thinner skim of mud, drenched by flurries of icy drops shaken from the atmosphere by a passing moan and a crash, breathing air heavy with a sweet, horrible, penetrating odor—such was the world as it existed for an hour one night, while I and the Commandant of *Douaumont* wandered about completely lost, on the top of his own fort. We finally stumbled on the little grated opening through which the lookout peered unceasingly over the landscape of mud. The mist lifted and we rediscovered the cave-like entrance, watched for a moment the ominous golden dumb-bells rising from the premier ligne, scraped our boots on a German helmet and went down again into the strangest sanctuary in the world.

This was the vision which flashed through my mind as I began vigil at an enormous nest of

Attas—the leaf-cutting ants of the British Guiana jungle. In front of me was a glade, about thirty feet across, devoid of green growth, and filled with a great irregular expanse of earth and mud. Relative to the height of the Attas, my six feet must seem a good half mile, and from this height I looked down and saw again the same inconceivably sticky clay of France. There were the rain-washed gullies, the half-roofed entrances to the vast underground fortresses, clean-swept, perfect roads, as efficient as the arteries of Verdun, flapping dead leaves like the omnipresent, worn-out scare-crows of camouflage, and over in one corner, to complete the simile, were a dozen shell-holes, the homes of voracious ant-lions, which, for passing insects, were unexploded mines, set at hair trigger.

My Atta city was only two hundred feet away from the laboratory, in fairly high jungle, within sound of the dinner triangle, and of the lapping waves on the Mazaruni shore. To sit near by and concentrate solely upon the doings of these ant people, was as easy as watching a single circus ring of performing elephants, while two more rings, a maze of trapezes, a race track and side-shows were in full swing. The jungle around me teemed with interesting happenings and distracting sights and sounds. The very last time I visited the nest and became absorbed in a line of incoming ants, I heard the shrill squeaking of an angry hummingbird overhead. I looked up, and there, ten feet above, was a furry tamandua anteater slowly climbing a straight purpleheart trunk, while around and around his head buzzed and swore the little fury—a pinch of cinnamon feathers, ablaze with rage. The curved claws of the unheeding anteater fitted around the trunk and the strong prehensile tail flattened against the bark, so that the creature seemed to put forth no more exertion than if walking along a fallen log. Now and then it stopped and daintily picked at a bit of termite nest.

With such side-shows it was sometimes difficult to concentrate on the Attas. Yet they offered problems for years of study. The glade was a little world in itself, with visitors and tenants, comedy and tragedy, sounds and silences. It was an ant-made glade, with all new growths either choked by upflung, earthen hillocks, or leaves bitten off as soon as they appeared. The casual vistors were the most conspicuous, an occasional trogon swooping across—a glowing, feathered comet of emerald, azurite and gold; or, slowly drifting in and out among the vines and coming to rest with waving wings, a yellow and red spotted Ithomiid,—or was it a Heliconiid or a Danaiid?—with such bewildering models and marvelous mimics it was impossible to tell without capture and close examination. Giant, purple tarantula-hawks hummed past, scanning the leaves for their prey.

Another class of glade haunters were those who came strictly on business,—plasterers and sculptors, who found wet clay ready to their needs. Great golden and rufous bees blundered down and gouged out bucketsful of mud; while slender-bodied, dainty, ebony wasps, after much fastidious picking of place, would detach a tiny bit of the whitest clay, place it in their snuff-box holder, clean their feet and antennæ, run their rapier in and out and delicately take to wing.

Little black trigonid bees had their special quarry, a small deep valley in the midst of a waste of interlacing Bad Lands, on the side of a precipitous butte. Here they picked and shoveled to their hearts' content, plastering their thighs until their wings would hardly lift them. They braced their feet, whirred, lifted unevenly, and sank back with a jar. Then turning, they bit off a piece of ballast, and heaving it over the precipice, swung off on an even keel.

Close examination of some of the craters and volcanic-like cones revealed many species of ants, beetles and roaches searching for bits of food—the scavengers of this small world. But the most interesting were the actual parasites, flies of many colors and sizes, humming past like little planes and zeppelins over this hidden city, ready to drop a bomb in the form of an egg deposited on the refuse heaps or on the ants themselves. The explosion might come slowly, but it would be none the less deadly. Once I detected a hint of the complexity of the glade life—beautiful metallic green flies walking swiftly about on long legs, searching nervously, whose eggs would be deposited near those of other flies, their larvæ to feed upon the others—parasites upon parasites.

As I had resolutely put the doings of the tree-tops away from my consciousness, so now I forgot visitors and parasites, and armed myself for the excavation of this buried metropolis. I rubbed vaseline on my high boots, and about the tops bound a band of teased-out absorbent cotton. My pick and shovel I treated likewise, and thus I was comparatively insulated. Without precautions no living being could withstand the slow, implacable attack of disturbed Attas. At present I walked unmolested across the glade. The millions beneath my feet were as unconscious of my presence as they were of the breeze in the palm fronds overhead.

At the first deep shovel thrust, a slow-moving flood of reddish-brown began to pour forth from the crumbled earth—the outposts of the Atta Maxims moving upward to the attack. For a few seconds only workers of various sizes appeared, then an enormous head heaved upward and there came into the light of day the first Atta soldier. He was twice as large as a large worker and heavy in proportion. Instead of being drawn up into two spines, the top of his head was rounded, bald and shiny, and only at the back were the two spines visible, shifted downward. The front of the head was thickly clothed with golden hair, which hung down bang-like over a round, glistening, single, median eye. One by one, and then shoulder to shoulder, these Cyclopean Maxims lumbered forth to battle, and soon my boots were covered in spite of the grease, all sinking their mandibles deep into the leather.

When I unpacked these boots this year I found the heads and jaws of two Attas still firmly attached, relics of some forgotten foray of the preceding year. This mechanical, vise-like grip, wholly independent of life or death, is utilized by the Guiana Indians. In place of stitching up extensive wounds, a number of these giant Atta Maxims are collected, and their jaws applied to the edges of the skin, which are drawn together. The ants take hold, their bodies are snipped off, and the row of jaws remains until the wound is healed.

Over and around the out-pouring soldiers, the tiny workers ran and bit and chewed away at whatever they could reach. Dozens of ants made their way up to the cotton, but found the utmost difficulty in clambering over the loose fluff. Now and then, however, a needle-like nip at the back of my neck, showed that some pioneer of these shock troops had broken through, when I was thankful that Attas could only bite and not sting as well. At such a time as this, the greatest difference is apparent between these and the Eciton

army ants. The Eciton soldier with his long, curved scimitars and his swift, nervous movements, was to one of these great insects as a fighting d'Artagnan would be to an armored tank. The results were much the same however, —perfect efficiency.

I now dug swiftly and crashed with pick down through three feet of soil. The great entrance arteries of the nest branched and bifurcated, separated and anastomosed, while here and there were chambers varying in size from a cocoanut to a football. These were filled with what looked like soft grayish sponge covered with whitish mold, and these somber affairs were the *raison d'être* for all the leaf-cutting, the trails, the struggles through jungles, the constant battling against wind and rain and sun.

But the labors of the Attas are only renewed when a worker disappears down a hole with his hard-earned bit of leaf. He drops it and goes on his way. We do not know what this way is, but my guess is that he turns around and goes after another leaf. Whatever the nests of Attas possess, they are without recreation rooms. These sluggard-instructors do not know enough to take a vacation; their faces are fashioned for biting, but not for laughing or yawning. I once dabbed fifteen Mediums with a touch of white paint as they approached the nest, and within five minutes thirteen of them had emerged and started on the back track again.

The leaf is taken in charge by another Medium, hosts of whom are everywhere. Once after a spadeful, I placed my eye as close as possible to a small heap of green leaves, and around one oblong bit were five Mediums, each with a considerable amount of chewed and mumbled tissue in front of him. This is the only time I have ever succeeded in finding these ants actually at this work. The leaves are chewed thoroughly and built up into the sponge gardens, being used neither for thatch nor for food, but as fertilizer. And not for any strange subterranean berry or kernel or fruit, but for a fungus or mushroom. The spores sprout and proliferate rapidly, the gray mycelia covering the garden, and at the end of each thread is a little knobbed body filled with liquid. This forms the sole food of the ants in the nest, but a drop of honey placed by a busy trail will draw a circle of workers at any time—both Mediums and Minims, who surround it and drink their fill.

When the fungus garden is in full growth, the nest labors of the Minims begin, and until the knobbed bodies are actually ripe, they never cease to weed and to prune, thus killing off the multitude of other fungi and foreign organisms, and by pruning they keep their particular fungus growing, and prevent it from fructifying. The fungus of the Attas is a particular species with the resonant, Dunsanyesque name of *Rozites gongylophora.* It is quite unknown outside of the nests of these ants, and is as artificial as a banana.

Only in Calcutta bazaars at night, and in underground streets of Pekin, have I seen stranger beings than I unearthed in my Atta nest. Now and then there rolled out of a shovelful of earth, an unbelievably big and rotund Cicada larva—which in the course of time, whether in one or in seventeen years, would emerge as the great marbled winged *Cicada gigas,* spreading five inches from tip to tip. Small tarantulas, with beautiful wine-colored cephalothorax, made their home deep in the nest, guarded, perhaps, by their dense covering of hair; slender scorpions sidled out from the ruins. They were bare, with vulnerable joints, but they had the advantage of a pair of hands, and long, mobile arms, which could quickly and skilfully pluck an attacking ant from any part of their anatomy.

The strangest of all the tenants were the tiny, amber-colored roaches which clung frantically to the heads of the great soldier ants, or scurried over the tumultuous mounds, searching for a crevice sanctuary. They were funny, fat little beings, wholly blind, yet supremely conscious of the danger that threatened, and with only the single thought of getting below the surface as quickly as possible. The Attas had very few insect guests, but this cockroach is one which had made himself perfectly at home. Through century upon century he had become more and more specialized and adapted to Atta life, eyes slipping until they were no more than faint specks, legs and antennæ changing, gait becoming altered to whatever speed and carriage best suited little guests in big underground halls and galleries. He and his race had evolved unseen and unnoticed even by the Maxim policemen. But when nineteen hundred humanly historical years had passed, a man with a keen sense of fitness named him Little Friend of the Attas; and so for a few more years, until scientists give place to the next caste, *Attaphila* will, all unconsciously, bear a name.

Attaphilas have staked their whole gamble of existence on the continued possibility of guestship with the Attas. Although they lived near the fungus gardens they did not feed upon them, but gathered secretions from the armored skin of the giant soldiers, who apparently did not object, and showed no hostility to their diminutive masseurs. A summer boarder may be quite at home on a farm, and safe from all ordinary dangers, but he must keep out of the way of scythes and sickles if he chooses to haunt the hay-fields. And so Attaphila, snug and safe, deep in the heart of the nest, had to keep on the qui vive when the ant harvesters came to glean in the fungus gardens. Snip, snip, snip, on all sides in the musty darkness, the keen mandibles sheared the edible heads, and though the little Attaphilas dodged and ran, yet most of them, in course of time, lost part of an antenna or even a whole one.

Thus the Little Friend of the Leaf-cutters lives easily through his term of weeks or months, or perhaps even a year, and has nothing to fear for food or mate, or from enemies. But Attaphilas cannot all live in a single nest, and we realize that there must come a crisis, when they pass out into a strange world of terrible light and multitudes of foes. For these pampered, degenerate roaches to find another Atta nest unaided, would be inconceivable. In the big nest which I excavated I observed them on the back and heads not only of the large soldiers, but also of the queens which swarmed in one portion of the galleries; and indeed, of twelve queens, seven had roaches clinging to them. This has been noted also of a Brazilian species, and we suddenly realize what splendid sports these humble insects are. They resolutely prepare for their gamble—*l'aventure magnifique*—the slenderest fighting chance, and we are almost inclined to forget the irresponsible implacability of instinct, and cheer the little fellows for lining up on this forlorn hope. When the time comes, the queens leave, and are off up into the unheard-of sky as if an earthworm should soar with eagle's feathers; past the gauntlet of voracious flycatchers and hawks, to the millionth chance of meeting an acceptable male of the same species. After the mating, comes the solitary search for a suitable site, and only when the pitifully unfair gamble has been won by a single fortunate queen, does the Attaphila climb tremblingly down and accept what fate has sent. His ninety and nine fellows have met death in almost as many ways.

With the exception of these strange inmates there are very few tenants or guests in the nests of the Attas. Unlike the termites and Ecitons, who harbor a host of weird boarders, the leaf-cutters are able to keep their nest free from undesirables.

Once, far down in the nest, I came upon three young queens, recently emerged, slow and stupid, with wings dull and glazed, who crawled with awkward haste back into darkness. And again twelve winged females were grouped in one small chamber, restless and confused. This was the only glimpse I ever had of Atta royalty at home.

Good fortune was with me, however, on a memorable fifth of May, when returning from a monkey hunt in high jungle. As I came out into the edge of a clearing, a low humming attracted my attention. It was ventriloquial, and my ear refused to trace it. It sounded exactly like a great aerodrome far in the distance, with a score or more of planes tuning up. I chanced to see a large bee-like insect rising through the branches, and following back along its path, I suddenly perceived the rarest of sights—an Atta nest entrance boiling with the excitement of a flight of winged kings and queens. So engrossed were the ants that they paid no attention to me, and I was able to creep up close and kneel within two feet of the hole. The main nest was twenty feet away, and this was a special exit made for the occasion—a triumphal gateway erected far away from the humdrum leaf traffic.

The two-inch, arched hole led obliquely down into darkness, while brilliant sunshine illumined the earthen take-off and the surrounding mass of pink Mazaruni primroses. Up this corridor was coming, slowly, with dignity, as befitted the occasion, a pageant of royalty. The king males were more active, as they were smaller in size than the females, but they were veritable giants in comparison with the workers. The queens seemed like beings of another race, with their great bowed thorax supporting the folded wings, heads correspondingly large, with less jaw development, but greatly increased keenness of vision. In comparison with the Minims, these queens were as a human being one hundred feet in height.

I selected one large queen as she appeared and watched her closely. Slowly and with great effort she climbed the steep ascent into the blazing sunlight. Five tiny Minims were clinging to her body and wings, all scrubbing and cleaning as hard as they could. She chose a clear space, spread her wings, wide and flat, stood high upon her six legs and waited. I fairly shouted at this change, for slight though it was, it worked magic, and the queen Atta was a queen no more, but a miniature, straddle-legged aeroplane, pushed into position, and overrun by a crowd of mechanics, putting the finishing touches, tightening the wires, oiling every pliable crevice. A Medium came along, tugged at a leg and the obliging little plane lifted it for inspection. For three minutes this kept up, and then the plane became a queen and moved restlessly. Without warning, as if some irresponsible mechanic had turned the primed propellers, the four mighty wings whirred—and four Minims were hurled head over heels a foot away, snapped from their positions. The sound of the wings was almost too exact an imitation of the snarl of a starting plane—the comparison was absurd in its exactness of timbre and resonance. It was only a test, however, and the moment the queen became quiet the upset mechanics clambered back. They crawled beneath her, scraped her feet and antennæ, licked her eyes and jaws, and went over every shred of wing tissue. Then again she buzzed, this time sending only a single Minim sprawling. Again she stopped after lifting herself an inch, but immediately started up, and now rose rather unsteadily, but without pause, and slowly ascended above the nest and the primroses. Circling once, she passed through green leaves and glowing balls of fruit, into the blue sky.

Thus I followed the passing of one queen Atta into the jungle world, as far as human eyes would permit, and my mind returned to the mote which I had detected at an equally great height —the queen descending after her marriage—as isolated as she had started.

We have seen how the little blind roaches occasionally cling to an emerging queen and so are transplanted to a new nest. But the queen bears something far more valuable. More faithfully than ever virgin tended temple fires, each departing queen fills a little pouch in the lower part of her mouth with a pellet of the precious fungus, and here it is carefully guarded until the time comes for its propagation in the new nest.

When she has descended to earth and excavated a little chamber, she closes the entrance, and for forty days and nights labors at the founding of a new colony. She plants the little fungus cutting and tends it with the utmost solicitude. The care and feeding in her past life have stored within her the substance for vast numbers of eggs. Nine out of ten which she lays she eats to give her the strength to go on with her labors, and when the first larvæ emerge, they, too, are fed with surplus eggs. In time they pupate and at the end of six weeks the first workers—all tiny Minims—hatch. Small as they are, born in darkness, yet no education is needed. The Spirit of the Attas infuses them. Play and rest are the only things incomprehensible to them, and they take charge at once, of fungus, of excavation, of the care of the queen and eggs, the feeding of the larvæ, and as soon as the huskier Mediums appear, they break through into the upper world and one day the first bit of green leaf is carried down into the nest.

The queen rests. Henceforth, as far as we know, she becomes a mere egg-producing machine, fed mechanically by mechanical workers, the food transformed by physiological mechanics into yolk and then deposited. The aeroplane has become transformed into an incubator.

One wonders whether, throughout the long hours, weeks and months, in darkness which renders her eyes a mockery, there ever comes to her dull ganglion a flash of memory of The Day, of the rushing wind, the escape from pursuing puff-birds, the jungle stretching away for miles beneath, her mate, the cool tap of drops from a passing shower, the volplane to earth, and the obliteration of all save labor. Did she once look behind her, did she turn aside for a second, just to feel the cool silk of petals?

As we have seen, an Atta worker is a member of the most implacable labor-union in the world: he believes in a twenty-four hour day, no pay, no play, no rest—he is a cog in a machine-driven Good-for-the-greatest-number. After studying these beings for a week, one longs to go out and shout for kaisers and tsars, for selfishness and crime—anything as a relief from such terrible unthinking altruism. All Atta workers are born free and equal—which is well; and they remain so—which is what a Buddhist priest once called "gashang"—or so it sounded, and which he explained as a state where plants and animals and men were crystal-like in growth and existence. What a welcome sight it would be to see a Medium mount a bit of twig, antennæ a crowd of Minims about him, and start off on a foray of his own!

We may jeer or condemn the Attas for their hard-shell existence, but there comes to mind again and again, the wonder of it all. Are the hosts of little beings really responsible; have they not evolved into a pocket, a mental cul-de-sac, a swamping of individuality, pooling their personalities? And what is it they have gained—what pledge of success in food, in safety, in propagation? They are not separate entities, they have none of the freedom of action, of choice, of individuality of the solitary wasps. They are the somatic cells of the body politic, while deep within the nest are the guarded sexual cells—the winged kings and queens, which from time to time, exactly as in isolated organisms, are thrown off to propagate, and to found new nests. They, no less than the workers, are parts of something more subtle than the visible Attas and their material nest. Whether I go to the ant as sluggard, or myrmocologist, or accidentally, via Pterodactyl Pups, a day spent with them invariably leaves me with my whole being concentrated on this mysterious Atta Ego. Call it Vibration, Aura, Spirit of the nest, clothe ignorance in whatever term seems appropriate, we cannot deny its existence and power.

As with the Army ants, the flowing lines of leaf-cutters always brought to mind great arteries, filled with pulsating, tumbling corpuscles. When an obstruction appeared, as a fallen leaf, across the great sandy track, a dozen, or twenty or a hundred workers gathered—like leucocytes —and removed the interfering object. If I injured a worker who was about to enter the nest, I inoculated the Atta organism with a pernicious, foreign body. Even the victim himself was dimly aware of the law of fitness. Again and again he yielded to the call of the nest, only to turn aside at the last moment. From a normal link in the endless Atta chain, he had become an outcast—snapped at by every passing ant, self-banished, wandering off at nightfall to die somewhere in the wilderness of grass. When well, an Atta has relations but no friends, when ill, every jaw is against him.

As I write this seated at my laboratory table, by turning down my lamp and looking out, I can see the star dust of Orion's nebula, and without moving from my chair, Rigel, Sirius, Capella and Betelgeuze—the blue, white, yellow and red evolution of so-called lifeless cosmic matter. A few slides from the aquarium at my side reveal an evolutionary sequence to the heavenly host—the simplest of earthly organisms playing fast and loose with the borderland, not only of plants and animals, but of the one and of the many-celled. First a swimming lily, Stentor, a solitary animal bloom, twenty-five to the inch; Cothurnia, a double lily, and Gonium, with a quartet of cells clinging tremulously together, progressing unsteadily—materially toward the rim of my field of vision—in the evolution of earthly life toward sponges, peripatus, ants and man.

I was interrupted in my microcosmus just as it occurred to me that Chesterton would heartily approve of my approximation of Sirius and Stentor, of Capella and Cothurnia—the universe balanced. My attention was drawn from the atom Gonium—whose brave little spirit was striving to keep his foursome one—a primordial struggle toward unity of self and division of labor; my consciousness climbed the microscope tube and came to rest upon a slim glass of amber liquid on my laboratory table: a servant had brought a cocktail, for it was New Year's Eve. (Now the thought came that there were a number of worthy people who would also approve of this approximation!) I looked at the small spirituous luxury, and I thought of my friends in New York, and then of the Attas in front of the laboratory. With my electric flash I went out into the starlight, and found the usual hosts struggling nestward with their chlorophyll burdens, and rushing frantically out into the black jungle for more and yet more leaves. My mind swept back over evolution from star-dust to Kartabo compound, from Gonium to man, and to these leaf-cutting ants. And I wondered whether the Attas were any the better for being denied the stimulus of temptation, or whether I was any the worse for the opportunity of refusing a second glass. I went back into the house, and voiced a toast to tolerance, to temperance, and—to pterodactyls—and drank my cocktail.

IX

HAMMOCK NIGHTS

THERE is a great gulf between pancakes and truffles: an eternal, fixed, abysmal cañon. It is like the chasm between beds and hammocks. It is not to be denied and not to be traversed; for if pancakes with syrup are a necessary of life, then truffles with anything must be, by the very nature of things, a supreme and undisputed luxury, a regal food for royalty and the chosen of the earth. There cannot be a shadow of a doubt that these two are divided; and it is not alone a mere arbitrary division of poverty and riches as it would appear on the surface. It is an alienation brought about by profound and fundamental differences; for the gulf between them is that gulf which separates the prosaic, the ordinary, the commonplace, from all that is colored and enlivened by romance.

The romance of truffles endows the very word itself with a halo, an aristocratic halo full of mystery and suggestion. One remembers the hunters who must track their quarry through marshy and treacherous lands, and one cannot forget their confiding catspaw, that desolated pig, created only to be betrayed and robbed of the fungi of his labors. He is one of the pathetic characters of history, born to secret sorrow, victimized by those superior tastes which do not become his lowly station. Born to labor and to suffer, but not to eat. To this day he commands my sympathy; his ghost—lean, bourgeois, reproachful—looks out at me from every market-place in the world where the truffle proclaims his faithful service.

But the pancake is a pancake, nothing more. It is without inherent or artificial glamour; and this unfortunately, when you come right down to it, is true of food in general. For food, after all, is one of the lesser considerations; the connoisseur, the gourmet, even the gourmand, spends no more than four hours out of the day at his table. From the cycle, he may select four in which to eat; but whether he will or not, he must set aside seven of the twenty-four in which to sleep.

Sleeping, then, as opposed to eating, is of almost double importance, since it consumes nearly twice as much time—and time, in itself, is the most valuable thing in the world. Considered from this angle, it seems incredible that we have no connoisseurs of sleep. For we have none. Therefore it is with some temerity that I declare sleep to be one of the romances of existence, and not by any chance the simple necessary it is reputed to be.

However, this romance, in company with whatever is worthy, is not to be discovered without the proper labor. Life is not all truffles. Neither do they grow in modest back-yards to be picked of mornings by the maid-of-all-work. A mere bed, notwithstanding its magic camouflage of coverings, of canopy, of disguised pillows, of shining brass or fluted carven posts, is, pancake like, never surrounded by this aura of romance. No, it is hammock sleep which is the sweetest of all slumber. Not in the hideous, dyed affairs of our summer porches, with their miserable curved sticks to keep the strands apart, and their maddening creaks which grow in length and discord the higher one swings—but in a hammock woven by Carib Indians. An Indian hammock selected at random will not suffice; it must be a Carib and none other. For they, themselves, are part and parcel of the romance, since they are not alone a quaint and poetic people, but the direct descendants of those remote Americans who were the first to see the caravels of Columbus. Indeed, he paid the initial tribute to their skill, for in the diary of his first voyage he writes,—

"A great many Indians in canoes came to the ship to-day for the purpose of bartering their cotton, and *hamacas* or nets in which they sleep."

It is supposed that this name owes its being to the hamack tree, from the bark of which they were woven. However that may be, the modern hammock of these tropical Red Men is so light and so delicate in texture that during the day one may wear it as a sash, while at night it forms an incomparable couch.

But one does not drop off to sleep in this before a just and proper preparation. This presents complexities. First, the hammock must be slung with just the right amount of tautness; then, the novice must master the knack of winding himself in his blanket that he may slide gently into his aerial bed and rest at right angles to the tied ends, thus permitting the free side-meshes to curl up naturally over his feet and head. This cannot be taught. It is an art; and any art is one-tenth technique, and nine-tenths natural talent. However, it is possible to acquire a certain virtuosity, which, after all is said, is but pure mechanical skill as opposed to sheer genius. One might, perhaps, get a hint by watching the living chrysalid of a potential moon-moth wriggle back into its cocoon—but little is to be learned from human teaching. However, if, night after night, one observes his Indians, a certain instinctive knowledge will arise to aid and abet him in his task. Then, after his patient apprenticeship, he may reap as he has sowed. If it is to be disaster, it is as immediate as it is ignominious; but if success is to be his portion, then he is destined to rest, wholly relaxed, upon a couch encushioned and resilient beyond belief. He finds himself exalted and supreme above all mundane disturbances, with the treetops and the stars for his canopy, and the earth a shadowy floor far beneath. This gentle aerial support is distributed throughout hundreds of fine meshes, and the sole contact with the earth is through twin living boles, pulsing with swift running sap, whose lichened bark and moonlit foliage excel any tapestry of man's devising.

Perhaps it is atavistic—this desire to rest and swing in a hamaca. For these are not unlike the treetop couches of our arboreal ancestors, such a one as I have seen an orang-utan weave in a few minutes in the swaying crotch of a tree. At any rate, the hammock is not dependent upon four walls, upon rooms and houses, and it partakes altogether of the wilderness. Its movement is æolian—yielding to every breath of air. It has even its own weird harmony—for I have often heard a low, whistling hum as the air rushed through the cordage mesh. In a sudden tropical gale every taut strand of my hamaca has seemed a separate, melodious, orchestral note, while I was buffeted to and fro, marking time to some rhythmic and reckless tune of the wind playing fortissimo on the woven strings about me. The climax of this musical outburst was not without a mild element of danger—sufficient to create that enviable state of mind wherein the sense of security and the knowledge that a minor catastrophe may perhaps be brought about are weighed one against the other.

Special, unexpected, and interesting minor dangers are also the province of the hamaca. Once, in the tropics, a great fruit fell on the elastic strands and bounced upon my body. There was an ominous swish of the air in the sweeping arc which this missile described, also a goodly shower of leaves; and since the fusillade took place at midnight, it was, all in all, a somewhat alarming visitation. However, there were no honorable scars to mark its advent; and what is more important, from all my hundreds of hammock nights, I have no other memory of any actual or threatened danger which was not due to human carelessness or stupidity. It is true that once, in another continent, by the light of a campfire, I saw the long, liana-like body of a harmless tree-snake wind down from one of my fronded bed-posts and, like a living woof following its shuttle, weave a passing pattern of emerald through the pale meshes. But this heralded no harm, for the poisonous reptiles of that region never climb; and so, since I was worn out by a hard day, I shut my eyes and slept neither better nor worse because of the transient confidence of a neighborly serpent.

As a matter of fact, the wilderness provides but few real perils, and in a hammock one is safely removed from these. One lies in a stratum above all damp and chill of the ground, beyond the reach of crawling tick and looping leech; and with an enveloping *mosquitaro,* or mosquito shirt, as the Venezuelans call it, one is fortified even in the worst haunts of these most disturbing of all pests.

Once my ring rope slipped and the hammock settled, but not enough to wake me up and force me to set it to rights. I was aware that something had gone wrong, but, half asleep, I preferred to leave the matter in the lap of the gods. Later, as a result, I was awakened several times by the patting of tiny paws against my body, as small jungle-folk, standing on their hind-legs, essayed to solve the mystery of the swaying, silent, bulging affair directly overhead. I was unlike any tree or branch or liana which had come their way before; I do not doubt that they thought me some new kind of ant-nest, since these structures are alike only as their purpose in life is identical—for they express every possible variation in shape, size, color, design, and position. As for their curiosity, I could make no complaint, for, at best, my visitors could not be so inquisitive as I, inasmuch as I had crossed one ocean and two continents with no greater object than to pry into their personal and civic affairs as well as those of their neighbors. To say nothing of their environment and other matters.

That my rope slipped was the direct result of my own inefficiency. The hammock protects one from the dangers of the outside world, but like any man-made structure, it shows evidences of those imperfections which are part and parcel of human nature, and serve, no doubt, to make it interesting. But one may at least strive for perfection by being careful. Therefore tie the ropes of your hammock yourself, or examine and test the job done for you. The master of hammocks makes a knot the name of which I do not know—I cannot so much as describe it. But I would like to twist it again—two quick turns, a push and a pull; then, the greater the strain put upon it, the greater its resistance.

This trustworthiness commands respect and admiration, but it is in the morning that one feels the glow of real gratitude; for, in striking camp at dawn, one has but to give a single jerk and the rope is straightened out, without so much as a second's delay. It is the tying, however, which must be well done—this I learned from bitter experience.

It was one morning, years ago, but the memory of it is with me still, vivid and painful. One of the party had left her hammock, which was tied securely since she was skilful in such matters, to sit down and rest in another, belonging to a servant. This was slung at one end of a high, tropical porch, which was without the railing that surrounds the more pretentious verandahs of civilization, so that the hammock swung free, first over the rough flooring, then a little out over the yard itself. A rope slipped, the faulty knot gave way, and she fell backward—a seven-foot fall with no support of any kind by which she might save herself. A broken wrist was the price she had to pay for another's carelessness—a broken wrist which, in civilization, is perhaps, one of the lesser tragedies; but this was in the very heart of the Guiana wilderness. Many hours from ether and surgical skill, such an accident assumes alarming proportions. Therefore, I repeat my warning: tie your knots or examine them.

It is true, that, when all is said and done, a dweller in hammocks may bring upon himself any number of diverse dangers of a character never described in books or imagined in fiction. A fellow naturalist of mine never lost an opportunity to set innumerable traps for the lesser jungle-folk, such as mice and opossums, all of which he religiously measured and skinned, so that each, in its death, should add its mite to human knowledge. As a fisherman runs out set lines, so would he place his traps in a circle under his hammock, using a cord to tie each and every one to the meshes. This done, it was his custom to lie at ease and wait for the click below which would usher in a new specimen,—perhaps a new species,—to be lifted up, removed, and safely cached until morning. This strategic method served a double purpose: it conserved natural energy, and it protected the catch. For if the traps were set in the jungle and trustfully confided to its care until the break of day, the ants would leave a beautifully cleaned skeleton, intact, all unnecessarily entrapped.

Now it happened that once, when he had set his nocturnal traps, he straightway went to sleep in the midst of all the small jungle people who were calling for mates and new life, so that he did not hear the click which was to warn him that another little beast of fur had come unawares upon his death. But he heard, suddenly, a disturbance in the low ferns beneath his hammock. He reached over and caught hold of one of the cords, finding the attendant trap heavy with prey. He was on the point of feeling his way to the trap itself, when instead, by some subconscious prompting, he reached over and snapped on his flashlight. And there before him, hanging in mid-air, striking viciously at his fingers which were just beyond its reach, was a young fer-de-lance—one of the deadliest of tropical serpents. His nerves gave way, and with a crash the trap fell to the ground where he could hear it stirring and thrashing about among the dead leaves. This ominous rustling did not encourage sleep; he lay there for a long time listening,—and every minute is longer in the darkness,—while his hammock quivered and trembled with the reaction.

Guided by this, I might enter into a new field of naturalizing and say to those who might, in excitement, be tempted to do otherwise, "Look at your traps before lifting them." But my audience would be too limited; I will refrain from so doing.

It is true that this brief experience might be looked upon as one illustration of the perils of the wilderness, since it is not customary for the fer-de-lance to frequent the city and the town. But this would give rise to a footless argument, leading nowhere. For danger is everywhere—it lurks in every shadow and is hidden in the bright sunlight, it is the uninvited guest, the invisible pedestrian who walks beside you in the crowded street ceaselessly, without tiring. But even a fer-de-lance should rather add to the number of hammock devotees than diminish them; for the three feet or more of elevation is as good as so many miles between the two of you. And three miles from any serpent is sufficient.

It may be that the very word danger is subjected to a different interpretation in each one of our mental dictionaries. It is elastic, comprehensive. To some it may include whatever is terrible, terrifying; to others it may symbolize a worthy antagonist, one who throws down the gauntlet and asks no questions, but who will make a good and fair fight wherein advantage is neither taken nor given. I suppose, to be bitten by vampires would be thought a danger by many who have not graduated from the mattress of civilization to this cubiculum of the wilderness. This is due, in part, to an ignorance, which is to be condoned; and this ignorance, in turn, is due to that lack of desire for a knowledge of new countries and new experiences, which lack is to be deplored and openly mourned. Many years ago, in Mexico, when I first entered the vampire zone, I was apprised of the fact by the clotted blood on my horse's neck in the early morning. In actually seeing this evidence, I experienced the diverse emotions of the discoverer, although as a matter of fact I had discovered nothing more than the verification of a scientific commonplace. It so happened that I had read, at one time, many conflicting statements of the workings of this aerial leech; therefore, finding myself in his native habitat, I went to all sorts of trouble to become a victim to his sorceries. The great toe is the favorite and stereotyped point of attack, we are told; so, in my hammock, my great toes were conscientiously exposed night after night, but not until a decade later was my curiosity satisfied.

I presume that this was a matter of ill luck, rather than a personal matter between the vampire and me. Therefore, as a direct result of this and like experiences, I have learned to make proper allowances for the whims of the Fates.

I have learned that it is their pleasure to deluge me with rainstorms at unpropitious moments, also to send me, with my hammock, to eminently desirable countries, which, however, are barren of trees and scourged of every respectable shrub. That the showers may not find me unprepared, I pack with my hamaca an extra length of rope, to be stretched taut from foot-post to head-post, that a tarpaulin or canvas may be slung over it. When a treeless country is presented to me in prospect, I have two stout stakes prepared, and I do not move forward without them.

It is a wonderful thing to see an experienced hammocker take his stakes, first one, then the other, and plunge them into the ground three or four times, measuring at one glance the exact distance and angle, and securing magically that mysterious "give" so essential to well-being and comfort. Any one can sink them like fence-posts, so that they stand deep and rigid, a reproach and an accusation; but it requires a particular skill to judge by the pull whether or not they will hold through the night and at the same time yield with gentle and supple swing to the least movement of the sleeper. A Carib knows, instantly, worthy and unworthy ground. I have seen an Indian sink his hamaca posts into sand with one swift, concentrated motion, mathematical in its precision and surety, so that he might enter at once into a peaceful night of tranquil and unbroken slumber, while I, a tenderfoot then, must needs beat my stakes down into the ground with tremendous energy, only to come to earth with a resounding thwack the moment I mounted my couch.

The Red Man made his comment, smiling: "Yellow earth, much squeeze." Which, being translated, informed me that the clayey ground I had chosen, hard though it seemed, was more like putty in that it would slip and slip with the prolonged pressure until the post fell inward and catastrophe crowned my endeavor.

So it follows that the hammock, in company with an adequate tarpaulin and two trustworthy stakes, will survive the heaviest downpour as well as the most arid and uncompromising desert. But since it is man-made, with finite limitations, nature is not without means to defeat its purpose. The hammock cannot cope with the cold—real cold, that is, not the sudden chill of tropical night which a blanket resists, but the cold of the north or of high altitudes. This is the realm of the sleeping-bag, the joy of which is another story. More than once I have had to use a hammock at high levels, since there was nothing else at hand; and the numbness of the Arctic was mine. Every mesh seemed to invite a separate draught. The winds of heaven—all four—played unceasingly upon me, and I became in due time a swaying mummy of ice. It was my delusion that I was a dead Indian cached aloft upon my arboreal bier—which is not a normal state of mind for the sleeping explorer.

Anything rather than this helpless surrender to the elements. Better the lowlands and that fantastic shroud, the mosquitaro. For even to wind one's self into this is an experience of note. It is ingenious, and called the mosquito shirt because of its general shape, which is as much like a shirt as anything else. A large round center covers the hammock, and two sleeves extend up the supporting strands and inclose the ends, being tied to the ring-ropes. If at sundown swarms of mosquitoes become unbearable, one retires into his netting funnel, and there disrobes. Clothes are rolled into a bundle and tied to the hammock, that one may close one's eyes reasonably confident that the supply will not be diminished by some small marauder. It is then that a miracle is enacted. For one is at last enabled, under these propitious circumstances, to achieve the impossible, to control and manipulate the void and the invisible, to obey that unforgotten advice of one's youth, "Oh, g'wan—crawl into a hole and pull the hole in after you!" At an early age, this unnatural advice held my mind, so that I devised innumerable means of verifying it; I was filled with a despair and longing whenever I met it anew. But it was an ambition appeased only in maturity. And this is the miracle of the tropics: climb up into the hamaca, and, at this altitude, draw in the hole of the mosquitaro funnel, making it fast with a single knot. It is done. One is at rest, and lying back, listens to the humming of all the mosquitos in the world, to be lulled to sleep by the sad, minor singing of their myriad wings. But though I have slung my hammock in many lands, on all the continents, I have few memories of netting nights. Usually, both in tropics and in tempered climes, one may boldly lie with face uncovered to the night.

And this brings us to the greatest joy of hammock life, admission to the secrets of the wilderness, initiation to new intimacies and subtleties of this kingdom, at once welcomed and delicately ignored as any honored guest should be. For this one must make unwonted demands upon one's nocturnal senses. From habit, perhaps, it is natural to lie with the eyes wide open, but with all the faculties concentrated on the two senses which bring impressions from the world of darkness—hearing and smell. In a jungle hut a loud cry from out of the black treetops now and then reaches the ear; in a tent the faint noises of the night outside are borne on the wind, and at times the silhouette of a passing animal moves slowly across the heavy cloth; but in a hamaca one is not thus set apart to be baffled by hidden mysteries—one is given the very point of view of the creatures who live and die in the open.

Through the meshes which press gently against one's face comes every sound which our human ears can distinguish and set apart from the silence—a silence which in itself is only a mirage of apparent soundlessness, a testimonial to the imperfection of our senses. The moaning and whining of some distant beast of prey is brought on the breeze to mingle with the silken swishing of the palm fronds overhead and the insistent chirping of many insects—a chirping so fine and shrill that it verges upon the very limits of our hearing. And these, combined, unified, are no more than the ground surge beneath the countless waves of sound. For the voice of the jungle is the voice of love, of hatred, of hope, of despair—and in the night-time, when the dominance of sense-activity shifts from eye to ear, from retina to nostril, it cries aloud its confidences to all the world. But the human mind is not equal to a true understanding of these; for in a tropical jungle the birds and the frogs, the beasts and the insects are sending out their messages so swiftly one upon the other, that the senses fail of their mission and only chaos and a great confusion are carried to the brain. The whirring of invisible wings and the movement of the wind in the low branches become one and the same: it is an epic, told in some strange tongue, an epic filled to overflowing with tragedy, with poetry and mystery. The cloth of this drama is woven from many-colored threads, for Nature is lavish with her pigment, reckless with life and death. She is generous because there is no need for her to be miserly. And in the darkness, I have heard the working of her will, translating as best I could.

In the darkness, I have at times heard the

tramping of many feet; in a land traversed only by Indian trails I have listened to an overloaded freight train toiling up a steep grade; I have heard the noise of distant battle and the cries of the victor and the vanquished. Hard by, among the trees, I have heard a woman seized, have heard her crying, pleading for mercy, have heard her choking and sobbing till the end came in a terrible, gasping sigh; and then, in the sudden silence, there was a movement and thrashing about in the topmost branches, and the flutter and whirr of great wings moving swiftly away from me into the heart of the jungle—the only clue to the author of this vocal tragedy. Once, a Pan of the woods tuned up his pipes—striking a false note now and then, as if it were his whim to appear no more than the veriest amateur; then suddenly, with the full liquid sweetness of his reeds, bursting into a strain so wonderful, so silvery clear, that I lay with mouth open to still the beating of blood in my ears, hardly breathing, that I might catch every vibration of his song. When the last note died away, there was utter stillness about me for an instant—nothing stirred, nothing moved; the wind seemed to have forsaken the leaves. From a great distance, as if he were going deeper into the woods, I heard him once more tuning up his pipes; but he did not play again.

Beside me, I heard the low voice of one of my natives murmuring, *"Muerte ha pasado."* My mind took up this phrase, repeating it, giving it the rhythm of Pan's song—a rhythm delicate, sustained, full of color and meaning in itself. I was ashamed that one of my kind could translate such sweet and poignant music into a superstition, could believe that it was the song of death,—the death that passes,—and not the voice of life. But it may have been that he was wiser in such matters than I; superstitions are many times no more than truth in masquerade. For I could call it by no name—whether bird or beast, creature of fur or feather or scale. And not for one, but for a thousand creatures within my hearing, any obscure nocturnal sound may have heralded the end of life. Song and death may go hand in hand, and such a song may be a beautiful one, unsung, unuttered until this moment when Nature demands the final payment for what she has given so lavishly. In the open, the dominant note is the call to a mate, and with it, that there may be color and form and contrast, there is that note of pure vocal exuberance which is beauty for beauty and for nothing else; but in this harmony there is sometimes the cry of a creature who has come upon death unawares, a creature who has perhaps been dumb all the days of his life, only to cry aloud this once for pity, for mercy, or for faith, in this hour of his extremity. Of all, the most terrible is the death-scream of a horse,—a cry of frightful timbre,—treasured, according to some secret law, until this dire instant when for him death indeed passes.

It was years ago that I heard the pipes of Pan; but one does not forget these mysteries of the jungle night: the sounds and scents and the dim, glimpsed ghosts which flit through the darkness and the deepest shadow mark a place for themselves in one's memory, which is not erased. I have lain in my hammock looking at a tapestry of green draped over a half-fallen tree, and then for a few minutes have turned to watch the bats flicker across a bit of sky visible through the dark branches. When I looked back again at the tapestry, although the dusk had only a moment before settled into the deeper blue of twilight, a score of great lustrous stars were shining there, making new patterns in the green drapery; for in this short time, the spectral blooms of the night had awakened and flooded my resting-place with their fragrance.

And these were but the first of the flowers; for when the brief tropic twilight is quenched, a new world is born. The leaves and blossoms of the day are at rest, and the birds and insects sleep. New blooms open, strange scents pour forth. Even our dull senses respond to these; for just as the eye is dimmed, so are the other senses quickened in the sudden night of the jungle. Nearby, so close that one can reach out and touch them, the pale Cereus moons expand, exhaling their sweetness, subtle breaths of fragrance calling for the very life of their race to the whirring hawkmoths. The tiny miller who, through the hours of glare has crouched beneath a leaf, flutters upward, and the trail of her perfume summons her mate perhaps half a mile down wind. The civet cat, stimulated by love or war, fills the glade with an odor so pungent that it seems as if the other senses must mark it.

Although there may seem not a breath of air in motion, yet the tide of scent is never still. One's moistened finger may reveal no cool side, since there is not the vestige of a breeze; but faint odors arrive, become stronger, and die away, or are wholly dissipated by an onrush of others, so musky or so sweet that one can almost taste them. These have their secret purposes, since Nature is not wasteful. If she creates beautiful things, it is to serve some ultimate end; it is her whim to walk in obscure paths, but her goal is fixed and immutable. However, her designs are hidden and not easy to decipher; at best, one achieves, not knowledge, but a few isolated facts.

Sport in a hammock might, by the casual thinker, be considered as limited to dreams of the hunt and chase. Yet I have found at my disposal a score of amusements. When the dusk has just settled down, and the little bats fill every glade in the forest, a box of beetles or grasshoppers—or even bits of chopped meat—offers the possibility of a new and neglected sport, in effect the inversion of baiting a school of fish. Toss a grasshopper into the air and he has only time to spread his wings for a parachute to earth, when a bat swoops past so quickly that the eyes refuse to see any single effort—but the grasshopper has vanished. As for the piece of meat, it is drawn like a magnet to the fierce little face. Once I tried the experiment of a bit of blunted bent wire on a long piece of thread, and at the very first cast I entangled a flutter-mouse and pulled him in. I was aghast when I saw what I had captured. A body hardly as large as that of a mouse was topped with the head of a fiend incarnate. Between his red puffed lips his teeth showed needle-sharp and ivory-white; his eyes were as evil as a caricature from *Simplicissimus,* and set deep in his head, while his ears and nose were monstrous with fold upon fold of skinny flaps. It was not a living face, but a mask of frightful mobility.

I set him free, deeming anything so ugly well worthy of life, if such could find sustenance among his fellows and win a mate for himself somewhere in this world. But he, for all his hideousness and unseemly mien, is not the vampire; the blood-sucking bat has won a mantle of deceit from the hands of Nature—a garb that gives him a modest and not unpleasing appearance, and makes it a difficult matter to distinguish him from his guileless confrères of our summer evenings.

But in the tropics,—the native land of the

hammock,—not only the mysteries of the night, but the affairs of the day may be legitimately investigated from this aerial point of view. It is a fetish of belief in hot countries that every unacclimatized white man must, sooner or later, succumb to that sacred custom, the siesta. In the cool of the day he may work vigorously, but this hour of rest is indispensable. To a healthful person, living a reasonable life, the siesta is sheer luxury. However, in camp, when the sun nears the zenith and the hush which settles over the jungle proclaims that most of the wild creatures are resting, one may swing one's hammock in the very heart of this primitive forest and straightway be admitted into a new province, where rare and unsuspected experiences are open to the wayfarer. This is not the province of sleep or dreams, where all things are possible and preëminently reasonable; for one does not go through sundry hardships and all manner of self-denial, only to be blindfolded on the very threshold of his ambition. No naturalist of a temperament which begrudges every unused hour will, for a moment, think of sleep under such conditions. It is not true that the rest and quiet are necessary to cool the Northern blood for active work in the afternoon, but the eye and the brain can combine relaxation with keenest attention.

In the northlands the difference in the temperature of the early dawn and high noon is so slight that the effect on birds and other creatures, as well as plants of all kinds, is not profound. But in the tropics a change takes place which is as pronounced as that brought about by day and night. Above all, the volume of sound becomes no more than a pianissimo melody; for the chorus of birds and insects dies away little by little with the increase of heat. There is something geometrical about this, something precise and fine in this working of a natural law —a law from which no living being is immune, for at length one unconsciously lies motionless, overcome by the warmth and this illusion of silence.

The swaying of the hammock sets in motion a cool breeze, and lying at full length, one is admitted at high noon to a new domain which has no other portal but this. At this hour, the jungle shows few evidences of life, not a chirp of bird or song of insect, and no rustling of leaves in the heat which has descended so surely and so inevitably. But from hidden places and cool shadows come broken sounds and whisperings, which cover the gamut from insects to mammals and unite to make a drowsy and contented murmuring—a musical undertone of amity and goodwill. For pursuit and killing are at the lowest ebb, the stifling heat being the flag of truce in the world-wide struggle for life and food and mate—a struggle which halts for naught else, day or night.

Lying quietly, the confidence of every unconventional and adventurous wanderer will include your couch, since courage is a natural virtue when the spirit of friendliness is abroad in the land. I felt that I had acquired merit that eventful day when a pair of hummingbirds—thimblefuls of fluff with flaming breastplates and caps of gold—looked upon me with such favor that they made the strands of my hamaca their boudoir. I was not conscious of their designs upon me until I saw them whirring toward me, two bright, swiftly moving atoms, glowing like tiny meteors, humming like a very battalion of bees. They betook themselves to two chosen cords and, close together, settled themselves with no further demands upon existence. A hundred of them could have rested upon the pair of strands; even the dragon-flies which dashed past had a wider spread of wing; but for these two there were a myriad glistening featherlets to be oiled and arranged, two pairs of slender wings to be whipped clean of every speck of dust, two delicate, sharp bills to be wiped again and again and cleared of microscopic drops of nectar. Then—like the great eagles roosting high overhead in the clefts of the mountainside—these mites of birds must needs tuck their heads beneath their wings for sleep; thus we three rested in the violent heat.

On other days, in Borneo, weaver birds have brought dried grasses and woven them into the fabric of my hammock, making me indeed feel that my couch was a part of the wilderness. At times, some of the larger birds have crept close to my glade, to sleep in the shadows of the low jungle-growth. But these were, one and all, timid folk, politely incurious, with evident respect for the rights of the individual. But once, some others of a ruder and more barbaric temperament advanced upon me unawares, and found me unprepared for their coming. I was dozing quietly, glad to escape for an instant the insistent screaming of a cicada which seemed to have gone mad in the heat, when a low rustling caught my ear—a sound of moving leaves without wind; the voice of a breeze in the midst of breathless heat. There was in it something sinister and foreboding. I leaned over the edge of my hammock, and saw coming toward me, in a broad, irregular front, a great army of ants, battalion after battalion of them flowing like a sea of living motes over twigs and leaves and stems. I knew the danger and I half sat up, prepared to roll out and walk to one side. Then I gaged my supporting strands; tested them until they vibrated and hummed, and lay back, watching, to see what would come about. I knew that no creature in the world could stay in the path of this horde and live. To kill an insect or a great bird would require only a few minutes, and the death of a jaguar or a tapir would mean only a few more. Against this attack, claws, teeth, poison-fangs would be idle weapons.

In the van fled a cloud of terrified insects—those gifted with flight to wing their way far off, while the humbler ones went running headlong, their legs, four, six, or a hundred, making the swiftest pace vouchsafed them. There were foolish folk who climbed up low ferns, achieving the swaying, topmost fronds only to be trailed by the savage ants and brought down to instant death.

Even the winged ones were not immune, for if they hesitated a second, an ant would seize upon them, and, although carried into the air, would not loosen his grip, but cling to them, obstruct their flight, and perhaps bring them to earth in the heart of the jungle, where, cut off from their kind, the single combat would be waged to the death. From where I watched, I saw massacres innumerable; terrible battles in which some creature—a giant beside an ant—fought for his life, crushing to death scores of the enemy before giving up.

They were a merciless army and their number was countless, with host upon host following close on each other's heels. A horde of warriors found a bird in my game-bag, and left of it hardly a feather. I wondered whether they would discover me, and they did, though I think it was more by accident than by intention. Nevertheless a half-dozen ants appeared on the foot-strands, nervously twiddling their antennæ in my direction. Their appraisal was brief; with no more than a second's delay they started toward me. I waited until they were well on their way, then vigorously twanged the cords under them harpwise, sending all the scouts into mid-air and headlong down among their fellows. So far as I know, this was a revolutionary maneuver in military tactics, comparable only to the explosion of a set mine. But even so, when the last of this brigade had gone on their menacing, pitiless way, and the danger had passed to a new province, I could not help thinking of the certain, inexorable fate of a man who, unable to move from his hammock or to make any defense, should be thus exposed to their attack. There could be no help for him if but one of this great host should scent him out and carry the word back to the rank and file.

It was after this army had been lost in the black shadows of the forest floor, that I remembered those others who had come with them—those attendant birds of prey who profit by the evil work of this legion. For, hovering over them, sometimes a little in advance, there had been a flying squadron of ant-birds and others which had come to feed, not on the ants, but on the insects which had been frightened into flight. At one time, three of these dropped down to perch on my hammock, nervous, watchful, and alert, waiting but a moment before darting after some ill-fated moth or grasshopper which, in its great panic, had escaped one danger only to fall an easy victim to another. For a little while, the twittering and chirping of these camp-followers, these feathered profiteers, was brought back to me on the wind; and when it had died away, I took up my work again in a glade in which no voice of insect reached my ears. The hunting ants had done their work thoroughly.

And so it comes about that by day or by night the hammock carries with it its own reward to those who have learned but one thing—that there is a chasm between pancakes and truffles. It is an open door to a new land which does not fail of its promise, a land in which the prosaic, the ordinary, the everyday have no place, since they have been shouldered out, dethroned, by a new and competent perspective. The god of hammocks is unfailingly kind, just, and generous to those who have found pancakes wanting and have discovered by inspiration, or what-not, that truffles do not grow in back-yards to be served at early breakfast by the maid-of-all-work. Which proves, I believe, that a mere bed may be a block in the path of philosophy, a commonplace, and that truffles and hammocks—hammocks unquestionably—are twin doors to the land of romance.

The swayer in hammocks may find amusement and may enrich science by his record of observations; his memory will be more vivid, his caste the worthier, for the intimacy with wild things achieved when swinging between earth and sky, unfettered by mattress or roof.

X

A TROPIC GARDEN

TAKE an automobile and into it pile a superman, a great evolutionist, an artist, an ornithologist, a poet, a botanist, a photographer, a musician, an author, adorable youngsters of fifteen, and a tired business man, and within half an hour I shall have drawn from them superlatives of appreciation, each after his own method of emotional expression—whether a flood of exclamations, or silence. This is no light boast, for at one time or another, I have done all this, but in only one place—the Botanical Gardens of Georgetown, British Guiana. As I hold it sacrilege to think of dying without again seeing the Taj Mahal, or the Hills from Darjeeling, so something of ethics seems involved in my soul's necessity of again watching the homing of the herons in these tropic gardens at evening.

In the busy, unlovely streets of the waterfront of Georgetown, one is often jostled; in the markets, it is often difficult at times to make one's way; but in the gardens a solitary laborer grubs among the roots, a coolie woman swings by with a bundle of grass on her head, or, in the late afternoon, an occasional motor whirrs past. Mankind seems almost an interloper, rather than architect and owner of these wonder-gardens. His presence is due far more often to business, his transit marked by speed, than the slow walking or loitering which real appreciation demands.

A guide-book will doubtless give the exact acreage, tell the mileage of excellent roads, record the date of establishment, and the number of species of palms and orchids. But it will have nothing to say of the marvels of the slow decay of a Victoria Regia leaf, or of the spiral descent of a white egret, or of the feelings which Roosevelt and I shared one evening, when four manatees rose beneath us. It was from a little curved Japanese bridge, and the next morning we were to start up-country to my jungle laboratory. There was not a ripple on the water, but here I chose to stand still and wait. After ten minutes of silence, I put a question and Roosevelt said, "I would willingly stand for two days to catch a good glimpse of a wild manatee." And St.

Francis heard, and, one after another, four great backs slowly heaved up; then an ill-formed head and an impossible mouth, with the unbelievable harelip, and before our eyes the sea-cows snorted and gamboled.

Again, four years later, I put my whole soul into a prayer for manatees, and again with success. During a few moments' interval of a tropical downpour, I stood on the same little bridge with Henry Fairfield Osborn. We had only half an hour left in the tropics; the steamer was on the point of sailing; what, in ten minutes, could be seen of tropical life! I stood helpless, waiting, hoping for anything which might show itself in this magic garden, where to-day the foliage was glistening malachite and the clouds a great flat bowl of oxidized silver.

The air brightened, and a tree leaning far across the water came into view. On its under side was a long silhouetted line of one and twenty little fish-eating bats, tiny spots of fur and skinny web, all so much alike that they might well have been one bat and twenty shadows.

A small crocodile broke water into air which for him held no moisture, looked at the bats, then at us, and slipped back into the world of crocodiles. A cackle arose, so shrill and sudden, that it seemed to have been the cause of the shower of drops from the palm-fronds; and then, on the great leaves of the Regia, which defy simile, we perceived the first feathered folk of this single tropical glimpse—spur-winged jacanas, whose rich rufus and cool lemon-yellow no dampness could deaden. With them were gallinules and small green herons, and across the pink mist of lotos blossoms just beyond, three egrets drew three lines of purest white—and vanished. It was not at all real, this onrush of bird and blossom revealed by the temporary erasing of the driven lines of gray rain.

Like a spendthrift in the midst of a winning game, I still watched eagerly and ungratefully for manatees. Kiskadees splashed rather than flew through the drenched air, an invisible black witch bubbled somewhere to herself, and a wren sang three notes and a trill which died out in a liquid gurgle. Then came another crocodile, and finally the manatees. Not only did they rise and splash and roll and indolently flick themselves with their great flippers, but they stood upright on their tails, like Alice's carpenter's companion, and one fondled its young as a water-mamma should. Then the largest stretched up as far as any manatee can ever leave the water, and caught and munched a drooping sprig of bamboo. Watching the great puffing lips, we again thought of walruses; but only a caterpillar could emulate that sideways mumbling—the strangest mouth of any mammal. But from behind, the rounded head, the shapely neck, the little baby manatee held carefully in the curve of a flipper, made legends of mermaids seem very reasonable; and if I had been an early *voyageur,* I should assuredly have had stories to tell of merkiddies as well. As we watched, the young one played about, slowly and deliberately, without frisk or gambol, but determinedly, intently, as if realizing its duty to an abstract conception of youth and warm-blooded mammalness.

The earth holds few breathing beings stranger than these manatees. Their life is a slow progression through muddy water from one bed of lilies or reeds to another. Every few minutes, day and night, year after year, they come to the surface for a lungful of the air which they must have, but in which they cannot live. In place of hands they have flippers, which paddle them leisurely along, which also serve to hold the infant manatee, and occasionally to scratch themselves when leeches irritate. The courtship of sea-cows, the qualities which appeal most to their dull minds, the way they protect the callow youngsters from voracious crocodiles, how or where they sleep—of all this we are ignorant. We belong to the same class, but the line between water and air is a no man's land which neither of us can pass for more than a few seconds.

When their big black hulks heaved slowly upward, it brought to my mind the huge glistening backs of elephants bathing in Indian streams; and this resemblance is not wholly fantastic. Not far from the oldest Egyptian ruins, excavations have brought to light ruins millions of years more ancient—the fossil bones of great creatures as strange as any that live in the realm of fairyland or fiction. Among them was revealed the ancestry of elephants, which was also that of manatees. Far back in geological times the tapir-like Moeritherium, which wandered through Eocene swamps, had within itself the prophecy of two diverse lines. One would gain great tusks and a long, mobile trunk and live its life in distant tropical jungles; and another branch was to sink still deeper into the swamp-water, where its hind-legs would weaken and vanish as it touched dry land less and less. And here to-day we watched a quartette of these manatees, living contented lives and breeding in the gardens of Georgetown.

The mist again drifted its skeins around leaf and branch, gray things became grayer, drops formed in mid-air and slipped slowly through other slower forming drops, and a moment later rain was falling gently. We went away, and to our mind's eye the manatees behind that gray curtain still munch bamboos, the spur-wings stretch their colorful wings cloudward, and the bubble-eyed crocodiles float intermittently between two watery zones.

To say that these are beautiful botanical gardens is like the statement that sunsets are admirable events. It is better to think of them as a setting, focusing about the greatest water-lily in the world, or, as we have seen, the strangest mammal; or as an exhibit of roots—roots as varied and as exquisite as a hall of famous sculpture; or as a wilderness of tapestry foliage, in texture from cobweb to burlap; or as a heaven-roofed, sun-furnaced greenhouse of blossoms, from the tiniest of dull-green orchids to the fifty-foot spike of taliput bloom. With this foundation of vegetation recall that the Demerara coast is a paradise for herons, egrets, bitterns, gallinules, jacanas, and hawks, and think of these trees and foliage, islands and marsh, as a nesting and roosting focus for hundreds of such birds. Thus, considering the gardens indirectly, one comes gradually to the realization of their wonderful character.

The Victoria Regia has one thing in common with a volcano—no amount of description or of colored plates prepares one for the plant itself. In analysis we recall its dimensions, colors, and form. Standing by a trench filled with its leaves and flowers, we discard the records of memory, and cleansing the senses of pre-impressions, begin anew. The marvel is for each of us, individually, an exception to evolution; it is a special creation, like all the rainbows seen in one's life —a thing to be reverently absorbed by sight, by scent, by touch, absorbed and realized without precedent or limit. Only ultimately do we find it necessary to adulterate this fine perception with definitive words and phrases, and so attempt to register it for ourselves or others.

I have seen many wonderful sights from an automobile,—such as my first Boche barrage and the tree ferns of Martinique,—but none to compare with the joys of vision from prehistoric *tikka gharries,* ancient victorias, and aged hacks. It was from the low curves of these equine rickshaws that I first learned to love Paris and Calcutta and the water-lilies of Georgetown. One of the first rites which I perform upon returning to New York is to go to the Lafayette and, after dinner, brush aside the taxi men and hail a victoria. The last time I did this, my driver was so old that two fellow drivers, younger than he and yet grandfatherly, assisted him, one holding the horse and the other helping him to his seat. Slowly ascending Fifth Avenue close to the curb and on through Central Park is like no other experience. The vehicle is so low and open that all resemblance to bus or taxi is lost. Everything is seen from a new angle. One learns incidentally that there is a guild of cab-drivers—proud, restrained, jealous. A hundred cars rush by without notice. Suddenly we see the whip brought up in salute to the dingy green top-hat, and across the avenue we perceive another victoria. And we are thrilled at the discovery, as if we had unearthed a new codex of some ancient ritual.

And so, initiated by such precedent, I have found it a worthy thing to spend hours in decrepit cabs loitering along side roads in the Botanical Gardens, watching herons and crocodiles, lilies and manatees, from the rusty leather seats. At first the driver looked at me in astonishment as I photographed or watched or wrote; but later he attended to his horse, whispering strange things into its ears, and finally deserted me. My writing was punctuated by graceful flourishes, resulting from an occasional lurch of the vehicle as the horse stepped from one to another patch of luscious grass.

Like Fujiyama, the Victoria Regia changes from hour to hour, color-shifted, wind-swung, and the mechanism of the blossoms never ceasing. In northern greenhouses it is nursed by skilled gardeners, kept in indifferent vitality by artificial heat and ventilation, with gaged light and selected water; here it was a rank growth, in its natural home, and here we knew of its antiquity from birds whose toes had been molded through scores of centuries to tread its great leaves.

In the cool fragrance of early morning, with the sun low across the water, the leaves appeared like huge, milky-white platters, with now and then little dancing silhouettes running over them. In another slant of light they seemed atolls scattered thickly through a dark, quiet sea, with new-blown flowers filling the whole air with slow-drifting perfume. Best of all, in late afternoon, the true colors came to the eye—six-foot circles of smooth emerald, with up-turned hem of rich wine-color. Each had a tell-tale cable lying along the surface, a score of leaves radiating from one deep hidden root.

Up through mud and black trench-water came the leaf, like a tiny fist of wrinkles, and day by day spread and uncurled, looking like the unwieldy paw of a kitten or cub. The keels and ribs covering the under-side increased in size and strength, and finally the great leaf was ironed out by the warm sun into a mighty sheet of smooth, emerald chlorophyll. Then, for a time, —no one has ever taken the trouble to find out how long,—it was at its best, swinging back and forth at its moorings with deep upright rim, a notch at one side revealing the almost invisible seam of the great lobes, and serving, also, as drainage outlet for excess of rain.

A young leaf occasionally came to grief by reaching the surface amid several large ones floating close together. Such a leaf expanded, as usual, but, like a beached boat, was gradually forced high and dry, hardening into a distorted shape and sinking only with the decay of the underlying leaves.

The deep crimson of the outside of the rim was merely a reflection tint, and vanished when the sun shone directly through; but the masses of sharp spines were very real, and quite efficient in repelling boarders. The leaf offered safe haven to any creature that could leap or fly to its surface; but its life would be short indeed if the casual whim of every baby crocodile or flipper of a young manatee met with no opposition.

Insects came from water and from air and called the floating leaf home, and, from now on, its surface was one of the most interesting and busy arenas in this tropical landscape.

In late September I spread my observation chair at the very edge of one of the dark tarns and watched the life on the leaves. Out at the center a fussy jacana was feeding with her two spindly-legged babies, while, still nearer, three scarlet-helmeted gallinules lumbered about, now and then tipping over a silvery and black infant which seemed puzzled as to which it should call parent. Here was a clear example, not only of the abundance of life in the tropics, but of the keen competition. The jacana invariably lays four eggs, and the gallinule, at this latitude, six or eight, yet only a fraction of the young had survived even to this tender age.

As I looked, a small crocodile rose, splashed, and sank, sending terror among the gallinules, but arousing the spur-wing jacana to a high pitch of anger. It left its young and flew directly to the widening circles and hovered, cackling loudly. These birds have ample ability to cope with the dangers which menace from beneath; but their fear was from above, and every passing heron, egret, or harmless hawk was given a quick scrutiny, with an instinctive crouch and half-spread wings.

But still the whole scene was peaceful; and as the sun grew warmer, young herons and egrets crawled out of their nests on the island a few yards away and preened their scanty plumage.

Kiskadees splashed and dipped along the margin of the water. Everywhere this species seems seized with an aquatic fervor, and in localities hundreds of miles apart I have seen them gradually desert their fly-catching for surface feeding, or often plunging, kingfisher-like, bodily beneath, to emerge with a small wriggling fish—another certain reflection of overpopulation and competition.

As I sat I heard a rustle behind me, and there, not eight feet away, narrow snout held high, one tiny foot lifted, was that furry fiend, Rikki-tikki. He was too quick for me, and dived into a small clump of undergrowth and bamboos. But I wanted a specimen of mongoose, and the artist offered to beat one end of the bush. Soon I saw the gray form undulating along, and as the rustling came nearer, he shot forth, moving in great bounds. I waited until he had covered half the distance to the next clump and rolled him over. Going back to my chair, I found that neither jacana, nor gallinules, nor herons had been disturbed by my shot.

While the introduction of the mongoose into Guiana was a very reckless, foolish act, yet he seems to be having a rather hard time of it, and with islands and lily-pads as havens, and waterways in every direction, Rikki is reduced chiefly to grasshoppers and such small game. He has spread along the entire coast, through the canefields and around the rice-swamps, and it will not be his fault if he does not eventually get a foothold in the jungle itself.

No month or day or hour fails to bring vital changes—tragedies and comedies—to the network of life of these tropical gardens; but as we drive along the broad paths of an afternoon, the quiet vistas show only waving palms, weaving vultures, and swooping kiskadees, with bursts of color from bougainvillea, flamboyant, and queen of the flowers. At certain times, however, the tide of visible change swelled into a veritable bore of life, gently and gradually, as quiet waters become troubled and then pass into the seething uproar of rapids. In late afternoon, when the long shadows of palms stretched their blue-black bars across the terra-cotta roads, the foliage of the green bamboo islands was dotted here and there with a scattering of young herons, white and blue and parti-colored. Idly watching them through glasses, I saw them sleepily preening their sprouting feathers, making ineffectual attempts at pecking one another, or else hunched in silent heron-dream. They were scarcely more alive than the creeping, hour-hand tendrils about them, mere double-stemmed, fluffy petaled blossoms, no more strange than the nearest vegetable blooms—the cannon-ball mystery, the sand-box puzzle, sinister orchids, and the false color-alarms of the white-bracted silver-leaf. Compared with these, perching herons are right and seemly fruit.

As I watched them I suddenly stiffened in sympathy, as I saw all vegetable sloth drop away and each bird become a detached individual, plucked by an electric emotion from the appearance of a thing of sap and fiber to a vital being of tingling nerves. I followed their united glance, and overhead there vibrated, lightly as a thistledown, the first incoming adult heron, swinging in from a day's fishing along the coast. It went on and vanished among the fronds of a distant island; but the calm had been broken, and through all the stems there ran a restless sense of anticipation, a zeitgeist of prophetic import. One felt that memory of past things was dimming, and content with present comfort was no longer dominant. It was the future to which both the baby herons and I were looking, and for them realization came quickly. The sun had sunk still lower, and great clouds had begun to spread their robes and choose their tints for the coming pageant.

And now the vanguard of the homing host appeared,—black dots against blue and white and salmon,—thin, gaunt forms with slow-moving wings which cut the air through half the sky. The little herons and I watched them come—first a single white egret, which spiralled down, just as I had many times seen the first returning Spad eddy downward to a cluster of great hump-backed hangars; then a trio of tricolored herons, and six little blues, and after that I lost count. It seemed as if these tiny islands were magnets drawing all the herons in the world.

Parrakeets whirl roostwards with machine-like synchronism of flight; geese wheel down in more or less regular formation; but these herons concentrated along straight lines, each describing its individual radius from the spot where it caught its last fish or shrimp to its nest or the particular branch on which it will spend the night. With a hemicircle of sufficient size, one might plot all of the hundreds upon hundreds of these radii, and each would represent a distinct line, if only a heron's width apart.

At the height of the evening's flight there were sometimes fifty herons in sight at once, beating steadily onward until almost overhead, when they put on brakes and dropped. Some, as the little egrets, were rather awkward; while the tricolors were the most skilful, sometimes nose-diving, with a sudden flattening out just in time to reach out and grasp a branch. Once or twice, when a fitful breeze blew at sunset, I had a magnificent exhibition of aeronautics. The birds came up-wind slowly, beating their way obliquely but steadily, long legs stretched out far behind the tail and swinging pendulum-like whenever a shift of ballast was needed. They apparently did not realize the unevenness of the wind, for when they backed air, ready to descend, a sudden gust would often undercut them and over they would go, legs, wings, and neck sprawling in mid-air. After one or two somersaults or a short, swift dive, they would right themselves, feathers on end, and frantically grasp at the first leaf or twig within reach. Panting, they looked helplessly around, reorientation coming gradually.

At each arrival, a hoarse chorus went up from

hungry throats, and every youngster within reach scrambled wildly forward, hopeful of a fish course. They received but scant courtesy and usually a vicious peck tumbled them off the branch. I saw a young bird fall to the water, and this mishap was from no attack, but due to his tripping over his own feet, the claws of one foot gripping those of the other in an insane clasp, which overbalanced him. He fell through a thin screen of vines and splashed half onto a small Regia leaf. With neck and wings he struggled to pull himself up, and had almost succeeded when heron and leaf sank slowly, and only the bare stem swung up again. A few bubbles led off in a silvery path toward deeper water, showing where a crocodile swam slowly off with his prey.

For a time the birds remained still, and then crept within the tangles, to their mates or nests, or quieted the clamor of the young with warm-storage fish. How each one knew its own offspring was beyond my ken, but on three separate evenings scattered through one week, I observed an individual, marked by a wing-gap of two lost feathers, come, within a quarter-hour of six o'clock, and feed a great awkward youngster which had lost a single feather from each wing. So there was no hit-or-miss method—no luck in the strongest birds taking toll from more than two of the returning parents.

Observing this vesper migration in different places, I began to see orderly segregation on a large scale. All the smaller herons dwelt together on certain islands in more or less social tolerance; and on adjoining trees, separated by only a few yards, scores of hawks concentrated and roosted, content with their snail diet, and wholly ignoring their neighbors. On the other side of the gardens, in aristocratic isolation, was a colony of stately American egrets, dainty and graceful. Their circumference of radiation was almost or quite a circle, for they preferred the ricefields for their daily hunting. Here the great birds, snowy white, with flowing aigrettes, and long, curving necks, settled with dignity, and here they slept and sat on their rough nests of sticks.

When the height of homing flight of the host of herons had passed, I noticed a new element of restlessness, and here and there among the foliage appeared dull-brown figures. There occurred the comic explanation of white herons who had crept deep among the branches, again emerging in house coat of drab! These were not the same, however, and the first glance through binoculars showed the thick-set, humped figures and huge, staring eyes of night herons.

As the last rays of the sun left the summit of the royal palms, something like the shadow of a heron flashed out and away, and then the import of these facts was impressed upon me. The egret, the night heron, the vampire—here were three types of organisms, characterizing the actions and reactions in nature. The islands were receiving and giving up. Their heart was becoming filled with the many day-feeding birds, and now the night-shift was leaving, and the very branch on which a night heron might have been dozing all day was now occupied, perhaps, by a sleeping egret. With eyes enlarged to gather together the scanty rays of light, the night herons were slipping away in the path of the vampires—both nocturnal, but unlike in all other ways. And I wondered if, in the very early morning, infant night herons would greet their returning parents; and if their callow young ever fell into the dark waters, what awful deathly alternates would night reveal; or were the slow-living crocodiles sleepless, with cruel eyes which never closed so soundly but that the splash of a young night heron brought instant response?

XI

THE BAY OF BUTTERFLIES

BUTTERFLIES doing strange things in very beautiful ways were in my mind when I sat down, but by the time my pen was uncapped my thoughts had shifted to rocks. The ink was refractory and a vigorous flick sent a shower of green drops over the sand on which I was sitting, and as I watched the ink settle into the absorbent quartz—the inversions of our grandmothers' blotters—I thought of what jolly things the lost ink might have been made to say about butterflies and rocks, if it could have flowed out slowly in curves and angles and dots over paper—for the things we might have done are always so much more worthy than those which we actually accomplish. When at last I began to write, a song came to my ears and my mind again looped backward. At least, there came from the very deeps of the water beyond the mangroves a low, metallic murmur; and my Stormouth says that in Icelandic *sangra* means to murmur. So what is a murmur in Iceland may very well be a song in Guiana. At any rate, my pen would have to do only with words of singing catfish; yet from butterflies to rock, to fish, all was logical looping —mental giant-swings which came as relaxation after hours of observation of unrelated sheer facts.

The singing cats, so my pen consented to write, had serenaded me while I crossed the Cuyuni in a canoe. There arose deep, liquid, vibrating sounds, such as those I now heard, deep and penetrating, as if from some submarine gong—a gong which could not be thought of as wet, for it had never been dry. As I stopped paddling the sound became absolute vibration, the canoe itself seemed to tremble, the paddle tingled in my hands. It was wholly detached; it came from whatever direction the ear sought it. Then, without dying out, it was reinforced by another sound, rhythmical, abrupt, twanging, filling the water and air with a slow measure on four notes. The water swirled beside the canoe, and a face appeared—a monstrous, complacent face, such as Böcklin would love—a face inhuman in possessing the quality of supreme contentment. Framed in the brown waters, the head of the great, grinning catfish rose, and slowly sank, leaving outlines discernible in ripples and bubbles with almost Cheshire persistency. One of my Indians, passing in his dugout, smiled at my peering down after the fish, and murmured, "Boom-boom."

Then came a day when one of these huge, amiable, living smiles blundered into our net, a smile a foot wide and six feet long, and even as he lay quietly awaiting what fate brought to great catfish, he sang, both theme and accompaniment. His whole being throbbed with the continuous deep drumming as the thin, silky walls of his swim-bladder vibrated in the depths of his body. The oxygen in the air was slowly killing him, and yet his swan song was possible because of an inner atmosphere so rich in this gas that it would be unbreathable by a creature of the land. Nerve and muscle, special expanse of circling bones, swim-bladder and its tenuous gas—all these combined to produce the aquatic harmony. But as if to load this contented being with largesse of apparently useless abilities, the two widespreading fin spines—the fins which correspond to our arms—were swiveled in rough-ridged cups at what might have been shoulders, and when moved back and forth the stridulation troubled all the water, and the air, too, with the muffled, twanging, *rip, rip, rip, rip.* The two spines were tuned separately, the right being a full tone lower, and the backward drawing of the bow gave a higher note than its forward reach. So, alternately, at a full second tempo, the four tones rose and fell, carrying out some strange Silurian theme: a muffled cadence of undertones, which, thrilled with the mystery of their author and cause, yet merged smoothly with the cosmic orchestra of wind and ripples and distant rain.

So the great, smooth, arching lift of granite rocks at our bungalow's shore, where the giant catfish sang, was ever afterward Boom-boom Point. And now I sat close by on the sand and strove to think anew of my butterflies, for they were the reason of my being there that brilliant October afternoon. But still my pen refused, hovering about the thing of ultimate interest as one leaves the most desired book to the last. For again the ear claimed dominance, and I listened to a new little refrain over my shoulder. I pictured a tiny sawhorse, and a midget who labored with might and main to cut through a never-ending stint of twigs. I chose to keep my image to the last, and did not move or look around, until there came the slightest of tugs at my knee, and into view clambered one of those beings who are so beautiful and bizarre that one almost thinks they should not be. My second singer was a beetle—an awkward, enormous, serious, brilliant beetle, with six-inch antennæ and great wing covers, which combined the hues of the royal robes of Queen Thi, tempered by thousands of years of silent darkness in the underground tombs at Sakhara, with the grace of curve and angle of equally ancient characters on the hill tombs of Fokien. On a background of olive ochre there blazed great splashes and characters of the red of jasper framed in black. Toward the front Nature had tried heavy black stippling, but it clouded the pattern and she had given it up in order that I might think of Egypt and Cathay.

But the thing which took the beetle quite out of a world of reasonable things was his forelegs. They were outrageous, and he seemed to think so, too, for they got in his way, and caught in wrong things and pulled him to one side. They were three times the length of his other limbs, spreading sideways a full thirteen inches, long, slender, beautifully sculptured, and forever reaching out in front for whatever long-armed beetles most desire. And his song, as he climbed over me, was squeaky and sawlike, and as he walked he doddered, head trembling as an old man's shakes in final acquiescence in the futility of life.

But in this great-armed beetle it was a nodding of necessity, a doddering of desire, the drawing of the bow across the strings in a hymn of hope which had begun in past time with the first stridulation of ancient insects. To-day the fiddling vibrations, the Song of the Beetle, reached out in all directions. To the majority of jungle ears it was only another note in the day's chorus: I saw it attract a flycatcher's attention, hold it a moment, and then lose it. To me it came as a vitally interesting tone of deep significance, for whatever emotions it might arouse in casual ears, its goal was another Great-armed Beetle, who might or might not come within its radius. With unquestioning search the fiddler clambered on and on, over me and over flowers and rocks, skirting the ripples and vanishing into a maelstrom of waving grass. Long after the last awkward lurch, there came back zizzing squeaks of perfect faith, and I hoped, as I passed beyond the periphery of sound, that instinct and desire might direct their rolling ball of vibrations toward the one whose ear, whether in antenna, or thorax or femoral tympanum had, through untold numbers of past lives, been attuned to its rhythm.

Two thousand miles north of where I sat, or ten million, five hundred and sixty thousand feet (for, like Bunker Bean's book-keeper, I sometimes like to think of things that way), I would look out of the window one morning in days to come, and thrill at the sight of falling flakes. The emotion would very probably be sentiment —the memory of wonderful northland snowstorms, of huge fires, of evenings with Roosevelt, when discussions always led to unknowable fields, when book after book yielded its phrase or sentence of pure gold thought. On one of the last of such evenings I found a forgotten joy-of-battle-speech of Huxley's, which stimulated two full days and four books re-read—while flakes swirled and invisible winds came swiftly around the eaves over the great trophies—*poussant des soupirs,*—we longing with our whole souls for an hour of talk with that splendid old fighting scientist.

These are thoughts which come at first-snow, thoughts humanly narrow and personal compared to the later delights of snow itself—crystals and tracks, the strangeness of freezing and the mystery of melting. And they recurred now because for days past I had idly watched scattered flurries of lemon-yellow and of orange butterflies drift past Kartabo. Down the two great Guiana rivers they came, steadily progressing, yet never hurrying; with zigzag flickering flight they barely cleared the trees and shrubs, and then skimmed the surface, vanishing when ripples caught the light, redoubled by reflection when the water lay quiet and polished. For month after month they passed, sometimes absent for days or weeks, but soon to be counted at earliest sunup, always arousing renewed curiosity, always bringing to mind the first flurry of winter.

We watch the autumn passing of birds with regret, but when the bluebirds warble their way southward we are cheered with the hope and the knowledge that some, at least, will return. Here, vast stretches of country, perhaps all Guiana, and how much of Brazil and Venezuela no one knows, poured forth a steady stream of yellow and orange butterflies. They were very beautiful and they danced and flickered in the sunlight, but this was no temporary shifting to a pleasanter clime or a land of more abundant flowers, but a migration in the grim old sense which Cicero loved, *non dubitat . . . migrare de vita.* No butterfly ever turned back, or circled again to the glade, with its yellow cassia blooms where he had spent his caterpillarhood. Nor did he fly toward the north star or the sunset, but between the two. Twelve years before, as I passed up the Essequibo and the Cuyuni, I noticed hundreds of yellow butterflies each true to his little compass variation of NNW.

There are times and places in Guiana where emigrating butterflies turn to the north or the south; sometimes for days at a time, but sooner or later the eddies straighten out, their little flotillas cease tacking, and all swing again NNW.

To-day the last of the migration stragglers of the year—perhaps the fiftieth great-grandsons of those others—held true to the Catopsilian lodestone.

My masculine pronouns are intentional, for of all the thousands and tens of thousands of migrants, all, as far as I know, were males. Catch a dozen yellows in a jungle glade and the sexes may be equal. But the irresistible maelstrom impels only the males. Whence they come or why they go is as utterly unknown to us as why the females are immune.

Once, from the deck of a steamer, far off the Guiana coast, I saw hosts of these same great saffron-wings flying well above the water, headed for the open sea. Behind them were sheltering fronds, nectar, soft winds, mates; before were corroding salt, rising waves, lowering clouds, a storm imminent. Their course was NNW, they sailed under sealed orders, their port was Death.

Looking out over the great expanse of the Mazaruni, the fluttering insects were usually rather evenly distributed, each with a few yards of clear space about it, but very rarely—I have seen it only twice—a new force became operative. Not only were the little volant beings siphoned up in untold numbers from their normal life of sleeping, feeding, dancing about their mates, but they were blindly poured into an invisible artery, down which they flowed in close association, *véritables corpuscules de papillons,* almost touching, forming a bending ribbon, winding its way seaward, with here and there a temporary fraying out of eddying wings. It seemed like a wayward cloud still stained with last night's sunset yellow, which had set out on its own path over rivers and jungles to join the sea mists beyond the uttermost trees.

Such a swarm seemed imbued with an ecstasy of travel which surpassed discomfort. Deep cloud shadows might settle down, but only dimmed the painted wings; under raindrops the ribbon sagged, the insects flying closer to the water. On the other hand, the scattered hosts of the more ordinary migrations, while they turned neither to the north nor to the west, yet fled at the advent of clouds and rain, seeking shelter under the nearest foliage. So much loitering was permitted, but with the coming of the sun again they must desert the pleasant feel of velvet leaves, the rain-washed odors of streaming blossoms, and set their antennæ unquestioningly upon the strange last turn of their wheel of life.

What crime of ancestors are they expiating? In some forgotten caterpillardom was an act committed, so terrible that it can never be known, except through the working out of the karma upon millions of butterflies? Or does there linger in the innumerable little ganglion minds a memory of long-lost Atlantis, so compelling to masculine Catopsilias that the supreme effort of their lives is an attempt to envisage it? "Absurd fancies, all," says our conscious entomological sense, and we agree and sweep them aside. And then quite as readily, more reasonable scientific theories fall asunder, and we are left at last alone with the butterflies, a vast ignorance, and a great unfulfilled desire to know what it all means.

On this October day the migration of the year had ceased. To my coarse senses the sunlight was of equal intensity, the breeze unchanged, the whole aspect the same—and yet something as intangible as thought, as impelling as gravitation, had ceased to operate. The tension once slackened, the butterflies took up their more usual lives. But what could I know of the meaning of "normal" in the life of a butterfly—I who boasted a miserable single pair of eyes and no greater number of legs, whose shoulders supported only shoulder blades, and whose youth was barren of caterpillarian memories!

As I have said, migration was at an end, yet here I had stumbled upon a Bay of Butterflies. No matter whether one's interest in life lay chiefly with ornithology, teetotalism, arrowheads, politics, botany, or finance, in this bay one's thoughts would be sure to be concentrated on butterflies. And no less interesting than the butterflies were their immediate surroundings. The day before, I had sat close by on a low boulder at the head of the tiny bay, with not a butterfly in sight. It occurred to me that my ancestor, Eryops, would have been perfectly at home, for in front of me were clumps of strange, carboniferous rushes, lacking leaves and grace, and sedges such as might be fashioned in an attempt to make plants out of green straw. Here and there an ancient jointed stem was in blossom, a pinnacle of white filaments, and hour after hour there came little brown trigonid visitors, stingless bees, whose nests were veritable museums of flower extracts—tubs of honey, hampers of pollen, barrels of ambrosia, hoarded in castles of wax. Scirpus-sedge or orchid, all was the same to them.

All odor evaded me until I had recourse to my usual olfactory crutch, placing the flower in a vial in the sunlight. Delicate indeed was the fragrance which did not yield itself to a few minutes of this distillation. As I removed the cork there gently arose the scent of thyme, and of rose petals long pressed between the leaves of old, old books—a scent memorable of days ancient to us, which in past lives of sedges would count but a moment. In an instant it passed, drowned in the following smell of bruised stem. But I had surprised the odor of this age-old growth, as evanescent as the faint sound of the breeze sifting through the cluster of leafless stalks. I felt certain that Eryops, although living among horserushes and ancient sedges, never smelled or listened to them, and a glow of satisfaction came over me at the thought that perhaps I represented an advance on this funny old forebear of mine; but then I thought of the little bees, drawn from afar by the scent, and I returned to my usual sense of human futility, which is always dominant in the presence of insect activities.

I leaned back, crowding into a crevice of rock, and strove to realize more deeply the kinship of these fine earth neighbors. Bone of my bone indeed they were, but their quiet dignity, their calmness in storm and sun, their poise, their disregard of all small, petty things, whether of mechanics, whether chemical or emotional—these were attributes to which I could only aspire, being the prerogatives of superiors.

These rocks, in particular, seemed of the very essence of earth. Three elements fought over them. The sand and soil from which they lifted their splendid heads sifted down, or was washed up, in vain effort to cover them. More subtly dead tree trunks fell upon them, returned to earth, and strove to encloak them. For six hours at a time the water claimed them, enveloping them slowly in a mantle of quicksilver, or surging over with rough waves. Algal spores took hold, desmids and diatoms swam in and settled down, little fish wandered in and out of the crevices, while large ones nosed at the entrances.

Then Mother Earth turned slowly onward; the moon, reaching down, beckoned with invisible fingers, and the air again entered this no man's land. Breezes whispered where a few moments before ripples had lapped; with the sun as ally, the last remaining pool vanished and there began the hours of aërial dominion. The most envied character of our lesser brethren is their faith. No matter how many hundreds of thousands of tides had ebbed and flowed, yet to-day every pinch of life which was blown or walked or fell or flew to the rocks during their brief respite from the waves, accepted the good dry surface without question.

Seeds and berries fell, and rolled into hollows rich in mulcted earth; parachutes, buoyed on thistle silk, sailed from distant jungle plants; every swirl of breeze brought spores of lichens and moss, and even the retreating water unwittingly aided, having transported hither and dropped a cargo of living things, from tiniest plant to seeds of mightiest mora. Though in the few allotted hours these might not sprout, but only quicken in their heart, yet blue-winged wasps made their faith more manifest, and worked with feverish haste to gather pellets of clay and fashion cells. I once saw even the beginning of storage—a green spider, which an hour later was swallowed by a passing fish instead of nourishing an infant wasp.

Spiders raised their meshes where shrimps had skipped, and flies hummed and were caught by singing jungle vireos, where armored catfish had passed an hour or two before.

So the elements struggled and the creatures

of each strove to fulfil their destiny, and for a little time the rocks and I wondered at it together.

In this little arena, floored with sand, dotted with rushes and balconied with boulders, many hundreds of butterflies were gathered. There were five species, all of the genius *Catopsilia,* but only three were easily distinguishable in life, the smaller, lemon yellow *statira,* and the larger, orange *argente* and *philea.* There was also *eubele,* the migrant, keeping rather to itself.

I took some pictures, then crept closer; more pictures and a nearer approach. Then suddenly all rose, and I felt as if I had shattered a wonderful painting. But the sand was a lodestone and drew them down. I slipped within a yard, squatted, and mentally became one of them. Silently, by dozens and scores, they flew around me, and soon they eclipsed the sand. They were so closely packed that their outstretched legs touched. There were two large patches, and a smaller area outlined by no boundary that I could detect. Yet when these were occupied the last comers alighted on top of the wings of their comrades, who resented neither the disturbance nor the weight. Two layers of butterflies crammed into small areas of sand in the midst of more sand, bounded by walls of empty air—this was a strange thing.

A little later, when I enthusiastically reported it to a professional lepidopterist he brushed it aside. "A common occurrence the world over, Rhopalocera gathered in damp places to drink." I, too, had observed apparently similar phenomena along icy streams in Sikhim, and around muddy buffalo-wallows in steaming Malay jungles. And I can recall many years ago, leaning far out of a New England buggy to watch clouds of little sulphurs flutter up from puddles beneath the creaking wheels.

The very fact that butterflies chose to drink in company is of intense interest, and to be envied as well by us humans who are temporarily denied that privilege. But in the Bay of Butterflies they were not drinking, nor during the several days when I watched them. One of the chosen patches of sand was close to the tide when I first saw them, and damp enough to appease the thirst of any butterfly. The other two were upon sand, parched by hours of direct tropical sun, and here the two layers were massed.

The insects alighted, facing in any direction,

but veered at once, heading upbreeze. Along the riverside of markets of tropical cities I have seen fleets of fishing boats crowded close together, their gay sails drying, while great ebony Neptunes brought ashore baskets of angel fish. This came to mind as I watched my flotillas of butterflies.

I leaned forward until my face was hardly a foot from the outliers, and these I learned to know as individuals. One sulphur had lost a bit of hind wing, and three times he flew away and returned to the same spot. Like most cripples, he was unamiable, and resented a close approach, pushing at the trespasser with a foreleg in a most unbutterfly-like way. Although I watched closely, I did not see a single tongue uncoiled for drinking. Only when a dense group became uneasy and pushed one another about were the tongue springs slightly loosened. Even the nervous antennæ were quiet after the insects had settled. They seemed to have achieved a Rhopaloceran Nirvana, content to rest motionless until caught up in the temporary whirlwinds of restlessness which now and then possessed them.

They came from all directions, swirling over the rocks, twisting through near-by brambles, and settling without a moment's hesitation. It was as though they had all been here many times before, a rendezvous which brooked not an instant's delay. From time to time some mass spirit troubled them, and, as one butterfly, the whole company took to wing. Close as they were when resting, they fairly buffeted one another in midair. Their wings, striking one another and my camera and face, made a strange little rustling, crisp and crackling whispers of sounds. As if a pile of Northern autumn leaves, fallen to earth, suddenly remembered days of greenness and humming bees, and strove to raise themselves again to the bare branches overhead.

Down came the butterflies again, brushing against my clothes and eyes and hands. All that I captured later were males, and most were fresh and newly emerged, with a scattering of dimmed wings, frayed at edges, who flew more slowly, with less vigor. Finally the lower patch was washed out by the rising tide, but not until the water actually reached them did the insects leave. I could trace with accuracy the exact reach of the last ripple to roll over the flat sand by the contour of the remaining outermost rank of insects.

On and on came the water, and soon I was forced to move, and the hundreds of butterflies in front of me. When the last one had left I went away, returning two hours later. It was then that I witnessed the most significant happening in the Bay of Butterflies—one which shook to the bottom the theory of my lepidopterist friend, together with my thoughtless use of the word normal. Over two feet of restless brown water covered the sand patches and rocked the scouring rushes. A few feet farther up the little bay the remaining sand was still exposed. Here were damp sand, sand dotted with rushes, and sand dry and white in the sun. About a hundred butterflies were in sight, some continually leaving, and others arriving. Individuals still dashed into sight and swooped downward. But not one attempted to alight on the exposed sand. There was fine, dry sand, warm to a butterfly's feet, or wet sand soaked with draughts of good Mazaruni water. But they passed this unheeding, and circled and fluttered in two swarms, as low as they dared, close to the surface of the water, exactly over the two patches of sand which had so drawn and held them or their brethren two hours before. Whatever the ultimate satisfaction may have been, the attraction was something transcending humidity, aridity, or immediate possibility of attainment. It was a definite cosmic point, a geographical focus, which, to my eyes and understanding, was unreasonable, unsuitable, and inexplicable.

As I watched the restless water and the butterflies striving to find a way down through it to the only desired patches of sand in the world, there arose a fine, thin humming, seeping up through the very waves, and I knew the singing catfish were following the tide shoreward. And as I considered my vast ignorance of what it all meant, of how little I could ever convey of the significance of the happenings in the Bay of Butterflies, I felt that it would have been far better for all of my green ink to have trickled down through the grains of sand.

XII

SEQUELS

TROPICAL midges of sorts live less than a day—sequoias have felt their sap quicken at the warmth of fifteen hundred springs. Somewhere between these extremes, we open our eyes, look about us for a time and close them again. Modern political geography and shifts of government give us Methusalistic feelings—but a glance at rocks or stars sends us shuddering among the other motes which glisten for a moment in the sunlight and then vanish.

We who strive for a little insight into evolution and the meaning of things as they are, forever long for a glimpse of things as they were. Here at my laboratory I wonder what the land was like before the dense mat of vegetation came to cover every rock and grain of sand, or how the rivers looked when first their waters trickled to the sea.

All our stories are of the middles of things,—without beginning or end; we scientists are plunged suddenly upon a cosmos in the full uproar of eons of precedent, unable to look ahead, while to look backward we must look down.

Exactly a year ago I spent two hours in a clearing in the jungle back of Kartabo laboratory, and let my eyes and ears have full swing.[1] Now in August of the succeeding year I came again to this clearing, and found it no more a clearing. Indeed so changed was it, that for weeks I had passed close by without a thought of the jungle meadow of the previous year, and now, what finally turned me aside from my usual trail, was a sound. Twelve months ago I wrote, "From the monotone of under-world sounds a strange little rasping detached itself, a reiterated, subdued scraping or picking. It carried my mind instantly to the throbbing theme of the Niebelungs, onomatopoetic of the little hammers forever busy in their underground work. I circled a small bush at my side, and found that the sound came from one of the branches near the top; so with my glasses I began a systematic search." This was as far as I ever got, for a flock of parrakeets exploded close at hand and blew the lesser sound out of mind. If I had stopped

[1] See page 34.

to guess I would probably have considered the author a longicorn beetle or some fiddling orthopter.

Now, a year later, I suddenly stopped twenty yards away, for at the end of the silvery cadence of a woodhewer, I heard the low, measured, toneless rhythm which instantly revived to mind every detail of the clearing. I was headed toward a distant palm frond beneath whose tip was a nest of Rufous Hermits, for I wished to see the two atoms of hummingbirds at the moment when they rolled from their *petit pois* egg-shells. I gave this up for the day and turned up the hill, where fifty feet away was the stump and bush near which I had sat and watched. Three times I went past the place before I could be certain, and even at the last I identified it only by the relative position of the giant tauroneero tree, in which I had shot many cotingas. The stump was there, a bit lower and more worn at the crevices, leaking sawdust like an overloved doll—but the low shrub had become a tall sapling, the weeds—vervain, boneset, velvet-leaf—all had been topped and killed off by dense-foliaged bushes and shrubs, which a year before had not raised a leaf above the meadow level. The old vistas were gone, the landscape had closed in, the wilderness was shutting down. Nature herself was "letting in the jungle." I felt like Rip Van Winkle, or even more alien, as if the passing of time had been accelerated and my longed-for leap had been accomplished, beyond the usual ken of mankind's earthly lease of senses.

All these astounding changes had come to pass through the heat and moisture of a tropical year, and under deliberate scientific calculation there was nothing unusual in the alteration. I remembered the remarkable growth of one of the laboratory bamboo shoots during the rainy season—twelve and a half feet in sixteen days, but that was a single stem like a blade of grass, whereas here the whole landscape was altered—new birds, new insects, branches, foliage, flowers, where twelve short months past, was open sky above low weeds.

In the hollow root on the beach, my band of crane-flies had danced for a thousand hours, but here was a sound which had apparently never ceased for more than a year—perhaps five thouhand hours of daylight. It was a low, penetrating, abruptly reiterated beat, occurring about once every second and a half, and distinctly audible a hundred feet away. The "low bush" from which it proceeded last year, was now a respectable sapling, and the source far out of reach overhead. I discovered a roundish mass among the leaves, and the first stroke of the ax sent the rhythm up to once a second, but did not alter the timbre. A few blows and the small trunk gave way and I fled for my life. But there was no angry buzzing and I came close. After a cessation of ten or fifteen seconds the sound began again, weaker but steady. The foliage was alive with small Azteca ants, but these were tenants of several small nests near by, and at the catastrophe overran everything.

The largest structure was the smooth carton nest of a wasp, a beautiful species, pale yellowish-red with wine-colored wings. Only once did an individual make an attempt to sting and even when my head was within six inches, the wasps rested quietly on the broken combs. By careful watching, I observed that many of the insects jerked the abdomen sharply downward, butting the comb or shell of smooth paper a forceful blow, and producing a very distinct noise. I could not at first see the mass of wasps which were giving forth the major rhythm, as they were hidden deep in the nest, but the fifty-odd wasps in sight kept perfect time, or occasionally an individual skipped one or two beats, coming in regularly on every alternate or every third beat. Where they were two or three deep, the uppermost wasps struck the insects below them with their abdomens in perfect rhythm with the nest beat. For half an hour the sound continued, then died down and was not heard again. The wasps dispersed during the night and the nest was deserted.

It reminded me of the telegraphing ants which I have often heard in Borneo, a remarkable sweeping roll, caused by the host of insects striking the leaves with their heads, and produced only when they are disturbed. It appeared to be of the nature of a warning signal, giving me opportunity to back away from the stinging legions which filled the thicket against which I pushed.

The rhythm of these wasps was very different. They were peaceable, not even resenting the devastation of their home, but always and always must the inexplicable beat, beat, beat, be kept up, serving some purpose quite hidden from me. During succeeding months I found two more nests, with similar fetish of sound vibrations, which led to their discovery. From one small nest, which fairly shook with the strength of their beats, I extracted a single wasp and placed him in a glass-topped, metal box. For three minutes he kept up the rhythmic beat. Then I began a more rapid tattoo on the bottom of the box, and the changed tempo confused him, so that he stopped at once, and would not tap again.

A few little Mazaruni daisies survived here and there, blossoming bravely, trying to believe that the shade was lessening, and not daily becoming more dense. But their leaves were losing heart, and paling in the scant light. Another six months and dead leaves and moss would have obliterated them, and the zone of brilliant flowers and gorgeous butterflies and birds would shift many feet into the air, with the tops of the trees as a new level.

As long as I remained by my stump my visitors were of the jungle. A yellow-bellied trogon came quite close, and sat as trogons do, very straight and stiff like a poorly mounted bird, watching passing flycatchers and me and the glimpses of sky. At first he rolled his little cuckoo-like notes, and his brown mate swooped up, saw me, shifted a few feet farther off and perched full of curiosity, craning her neck and looking first with one eye, then the other. Now the male began a content song. With all possible variations of his few and simple tones, on a low and very sweet timbre, he belied his unoscine perch in the tree of bird life, and sang to himself. Now and then he was drowned out by the shrilling of cicadas, but it was a delightful serenade, and he seemed to enjoy it as much as I did. A few days before, I had made a careful study of the syrinx of this bird, whom we may call rather euphoniously *Trogonurus curucui,* and had been struck by the simplicity both of muscles and bones. Now, having summoned his mate in regular accents, there followed this unexpected whisper song. It recalled similar melodies sung by pheasants and Himalayan partridges, usually after they had gone to roost.

Once the female swooped after an insect, and in the midst of one of the sweetest passages of the male trogon, a green grasshopper shifted his position. He was only two inches away from the singer, and all this time had been hidden by his chlorophyll-hued veil. And now the trogon fairly fell off the branch, seizing the insect almost before the tone died away. Swallowing it with considerable difficulty, the harmony was taken up again, a bit throaty for a few notes. Then the pair talked together in the usual trogon fashion, and the sudden shadow of a passing vulture, drew forth discordant cat calls, as both birds swooped from sight to avoid the fancied hawk.

A few minutes later the vocal seal of the jungle was uttered by a quadrille bird. When the notes of this wren are heard, I can never imagine open, blazing sunshine, or unobstructed blue sky. Like the call of the wood pewee, the wren's radiates coolness and shadowy quiet. No matter how tropic or breathless the jungle, when the flute-like notes arise they bring a feeling of freshness, they arouse a mental breeze, which cools one's thoughts, and, although there may be no water for miles, yet we can fairly hear the drip of cool drops falling from thick moss to pools below. First an octave of two notes of purest silver, then a varying strain of eight or ten notes, so sweet and powerful, so individual and meaningful that it might stand for some wonderful motif in a great opera. I shut my eyes, and I was deaf to all other sounds while the wren sang.

And as it dwelt on the last note of its phrase, a cicada took it up on the exact tone, and blended the two final notes into a slow vibration, beginning gently and rising with the crescendo of which only an insect, and especially a cicada, is master. Here was the eternal, hypnotic tom-tom rhythm of the East, grafted upon supreme Western opera. For a time my changed clearing became merely a sounding box for the most thrilling of jungle songs. I called the wren as well as I could, and he came nearer and nearer. The music rang out only a few yards away. Then he became suspicious, and after that each phrase was prefaced by typical wren scolding. He could not help but voice his emotions, and the harsh notes told plainly what he thought of my poor imitation. Then another feeling would dominate, and out of the maelstrom of harshness, of tumbled, volcanic vocalization would rise the pure silver stream of single notes.

The wren slipped away through the masses of fragrant Davilla blossoms, but his songs remained and are with me to this moment. And now I leaned back, lost my balance, and grasping the old stump for support, loosened a big piece of soft, mealy wood. In the hollow beneath, I saw a rainbow in the heart of the dead tree.

This rainbow was caused by a bug, and when we stop to think of it, this shows how little there is in a name. For when we say bug, or for that matter bogy or bugbear, we are garbling the sound which our very, very forefathers uttered when they saw a specter or hobgoblin. They said it *bugge* or even *bwg,* but then they were more afraid of specters in those days than we, who imprison will-o'-the-wisps in Very lights, and rub fox-fire on our watch faces. At any rate here was a bug who seemed to ill-deserve his name, although if the Niblelungs could fashion the Rheingold, why could not a bug conceive a rainbow?

Whenever a human, and especially a house-human thinks of bugs, she thinks unpleasantly and in superlatives. And it chances that evolution, or natural selection, or life's mechanism, or fate or a creator, has wrought them into form and function also in superlatives. Cicadas are supreme in longevity and noise. One of our northern species sucks in silent darkness for seventeen years, and then, for a single summer, breaks all American long-distance records for insect voices. To another group, known as Fulgorids, gigantic heads and streamers of wax have been allotted. Those possessing the former rejoice in the name of Lantern Flies, but they are at present unfaithful vestal bugs, though it is extremely doubtful if their wicks were ever trimmed or lighted. To see a big wax bug flying with trailing ribbons slowly from tree to tree in the jungle is to recall the streaming trains of a flock of peacocks on the wing.

The membracids most of all deserve the name of "bugges" for no elf or hobgoblin was ever more bizarre. Their legs and heads and bodies are small and aphid-like, but aloft there spring minarets and handles and towers and thorns and groups of hairy balls, out of all reason and sense. Only Stegosaurus and Triceratops bear comparison. Another group of five-sided bugs are the skunks and civet-cats among insects, guarding themselves from danger by an aura of obnoxious scent.

Not the least strange of this assemblage is the author of our rainbow in the stump. My awkwardness had broken into a hollow which opened to the light on the other side of the rotten bole. A vine had tendriled its way into the crevice where the little weaver of rainbows had found board and lodging. We may call him toad-hopper or spittle-bug, or as Fabre says," *Contentons-nous de Cicadelle, qui respecte le tympan."* Like all of its kindred, the Bubble Bug finds Nirvana in a sappy green stem. It has neither strong flight, nor sticky wax, thorny armature nor gas barrage, so it proceeds to fashion an armor of bubbles, a cuirass of liquid film. This, in brief, was the rainbow which caught my eye when I broke open the stump. Up to that moment no rainbow had existed, only a little light sifting through from the vine-clad side. But now a ray of sun shattered itself on the pile of bubbles, and sprayed itself out into a curved glory.

Bubble Bugs blow their froth only when immature, and their bodies are a distillery or home-brew of sorts. No matter what the color, or viscosity or chemical properties of sap, regardless of whether it flows in liana, shrub, or vine, yet the Bug's artesian product is clear, tasteless and wholly without the possibility of being blown into bubbles. When a large drop has collected, the tip of the abdomen encloses a retort of air, inserts this in the drop and forces it out. In some way an imponderable amount of oil or dissolved wax is extruded and mixed with the drop, an invisible shellac which toughens the bubble and gives it an astounding glutinous endurance. As long as the abdominal air-pump can be extended into the atmosphere, so long does the pile of bubbles grow until the insect is deep buried, and to penetrate this is as unpleasant an achievement for small marauders as to force a cobweb entanglement. I have draped a big pile of bubbles around the beak of an insect-eating bird, and watched it shake its head and wipe its beak in evident disgust at the clinging oily films. In the north we have the bits of fine white foam which we characteristically call frog-spittle, but these tropic relatives have bigger bellows and their covering is like the interfering mass of films which emerges from the soap-bubble bowl when a pipe is thrust beneath the surface and that delicious gurgling sound produced.

The most marvelous part of the whole thing is that the undistilled well which the Bubble Bug taps would often overwhelm it in an instant, either by the burning acidity of its composition, or the rubber coating of death into which it hardens in the air. Yet with this current of lava or vitriol, our Bug does three wonderful things, it distills sweet water for its present protective cell of bubbles, it draws purest nourishment for continual energy to run its bellows and pump, and simultaneously it fills its blood and tissues with a pungent flavor, which in the future will be a safeguard against the attacks of birds and lizards. Little by little its wings swell to full spread and strength, muscles are fashioned in its hind legs, which in time will shoot it through great distances of space, and pigment of the most brilliant yellow and black forms on its wing covers. When at last it shuts down its little still and creeps forth through the filmy veil, it is immature no longer, but a brilliant frog-hopper, sitting on the most conspicuous leaves, trusting by pigmental warning to advertise its inedibility, and watchful for a mate, so that the future may hold no dearth of Bubble Bugs.

On my first tramp each season in the tropical jungle, I see the legionary army ants hastening on their way to battle, and the leaf-cutters plodding along, with chlorophyll hods over their shoulders, exactly as they did last year, and the year preceding, and probably a hundred thousand years before that. The Colony Egos of army and leaf-cutters may quite reasonably be classified according to Kingdom. The former, with carnivorous, voracious, nervous, vitally active members, seems an intangible, animal-like organism; while the stolid, vegetarian, unemotional, weather-swung Attas, resemble the flowing sap of the food on which they subsist—vegetable.

Yet, whatever the simile, the net of unconscious precedent is too closely drawn, the mesh of instinct is too fine to hope for any initiative. This was manifested by the most significant and spectacular occurrence I have ever observed in the world of insects. One year and a half ago I studied and reported upon, a nest of Ecitons or army ants.[1] Now, eighteen months later, apparently the same army appeared and made a similar nest of their own bodies, in the identical spot near the door of the out-house, where I had found them before. Again we had to break up the temporary colony, and killed about three-quarters of the colony with various deadly chemicals.

In spite of all the tremendous slaughter, the Ecitons, in late afternoon, raided a small colony of Wasps-of-the-Painted-Nest. These little

[1] See page 58.

chaps construct a round, sub-leaf carton-home, as large as a golf ball, which carries out all the requirements of counter shading and of ruptive markings. The flattened, shadowed under surface was white, and most of the sloping walls dark brown, down which extended eight white lines, following the veins of the leaf overhead. The side close to the stem of the leaf, and consequently always in deep shadow, was pure white. The eaves catching high lights were black. All this marvelous merging with leaf tones went for naught when once an advance Eciton scout located the nest.

As the deadly mob approached, the wasplets themselves seemed to realize the futility of offering battle, and the entire colony of forty-four gathered in a forlorn group on a neighboring leaf, while their little castle was rifled—larvæ and pupæ torn from their cells and rushed down the stems to the chaos which was raging in Eciton's own home. The wasps could guard against optical discovery, but the blind Ecitons had senses which transcended vision, if not even scent.

Late that night, our lanterns showed the remnants of the Eciton army wandering aimlessly about, making near approach impossible, but apparently lacking any definite concerted action.

At six o'clock the following morning I started out for a swim, when at the foot of the laboratory steps I saw a swiftly-moving, broad line of army ants on safari, passing through the compound to the beach. I traced them back under the servants' quarters, through two clumps of bamboos to the out-house. Later I followed along the column down to the river sand, through a dense mass of underbrush, through a hollow log, up the bank, back through light jungle—to the out-house again, and on a large fallen log, a few feet beyond the spot where their nest had been, the ends of the circle *actually came together!* It was the most astonishing thing, and I had to verify it again and again before I could believe the evidence of my eyes. It was a strong column, six lines wide in many places, and the ants fully believed that they were on their way to a new home, for most were carrying eggs or larvæ, although many had food, including the larvæ of the Painted Nest Wasplets. For an hour at noon during heavy rain, the column weakened and almost disappeared, but when the sun returned, the lines rejoined, and the revolution of the vicious circle continued.

There were several places which made excellent points of observation, and here we watched and marveled. Careful measurement of the great circle showed a circumference of twelve hundred feet. We timed the laden Ecitons and found that they averaged two to two and three-quarter inches a second. So a given individual would complete the round in about two hours and a half. Many guests were plodding along with the ants, mostly staphylinids of which we secured five species, a brown histerid beetle, a tiny chalcid, and several Phorid flies, one of which was winged.

The fat Histerid beetle was most amusing, getting out of breath every few feet, and abruptly stopping to rest, turning around in its tracks, standing almost on its head, and allowing the swarm of ants to run up over it and jump off. Then on it would go again, keeping up the terrific speed of two and a half inches a second for another yard. Its color was identical with the Ecitons' armor, and when it folded up, nothing could harm it. Once a worker stopped and antennæd it suspiciously, but aside from this, it was accepted as one of the line of marchers. Along the same route came the tiny Phorid flies, wingless but swift as shadows, rushing from side to side, over ants, leaves, debris, impatient only at the slowness of the army.

All the afternoon the insane circle revolved; at midnight the hosts were still moving, the second morning many had weakened and dropped their burdens, and the general pace had very appreciably slackened. But still the blind grip of instinct held them. On, on, on they must go! Always before in their nomadic life there had been a goal—a sanctuary of hollow tree, snug heart of bamboos—surely this terrible grind must end somehow. In this crisis, even the Spirit of the Army was helpless. Along the normal paths of Eciton life he could inspire endless enthusiasm, illimitable energy, but here his material units were bound upon the wheel of their perfection of instinct. Through sun and cloud, day and night, hour after hour there was found no Eciton with individual initiative enough to turn aside an ant's breadth from the circle which he had traversed perhaps fifteen times: the masters of the jungle had become their own mental prey.

Fewer and fewer now came along the well worn path; burdens littered the line of march, like the arms and accoutrements thrown down by a retreating army. At last a scanty single line struggled past—tired, hopeless, bewildered, idiotic and thoughtless to the last. Then some half dead Eciton straggled from the circle along the beach, and threw the line behind him into confusion. The desperation of total exhaustion had accomplished what necessity and opportunity and normal life could not. Several others followed his scent instead of that leading back toward the out-house, and as an amoeba gradually flows into one of its own pseudopodia, so the forlorn hope of the great Eciton army passed slowly down the beach and on into the jungle. Would they die singly and in bewildered groups, or would the remnant draw together, and again guided by the super-mind of its Mentor lay the foundation of another army, and again come to nest in my out-house?

Thus was the ending still unfinished, the finale buried in the future—and in this we find the fascination of Nature and of Science. Who can be bored for a moment in the short existence vouchsafed us here; with dramatic beginnings barely hidden in the dust, with the excitement of every moment of the present, and with all of cosmic possibility lying just concealed in the future, whether of Betelgeuze, of Amoeba or—of ourselves? *Vogue la galère!*

APPENDIX OF SCIENTIFIC NAMES

Page	*Line*	
4	26	Moriche Oriole; *Icterus chrysocephalus* (Linné)
8	10	Toad; *Bufo guttatus* Schneid.
18	3	Bat; *Furipterus horrens* (F. Cuv.)
	4	Large Bats; *Vampyrus spectrum* (Linné)
	6	Vampire Bats; *Desmodus rotundus* (Geoff.)
22	5	Giant Catfish, Boom-boom; *Doras granulosus* Valen.
23	5	Kiskadee; *Pitangus s. sulphuratus* (Linné)
25	26	Parrakeets; *Touit batavica* (Bodd.)
	26	Great Black Orioles; *Ostinops d. decumanus* (Pall.)
26	5	House Wrens; *Troglodytes musculus clarus* Berl. and Hart
29	5	Coati-mundi; *Nasua n. nasua* (Linné)
32	2	Frog; *Phyllomedusa* sp.
34	18	Mazaruni Daisies; *Sipanea pratensis* Aubl.
	20	Button Weed; *Spermacoce* sp.
36	23	Melancholy Tyrant; *Tyrannus melancholicus satrapa* (Cab. and Hein.)
37	2	Monarch; *Anosia plexippus* (Linné)
38	7	Red-breasted Blue Chatterer; *Cotinga cotinga* Linné
	18	Yellow Papilio; *Papilio thoas* Linné
49	26	Parrakeets; *Touit batavica* (Bodd.)
52	3	Purple-throated Cotinga; *Cotinga cayana* (Linné)
53	15	Dark-breasted Mourner; *Lipaugus simplex* Licht.
54	26	Toucans; *Ramphastus vitellinus* Licht.
59	6	White-fronted Ant-bird; *Pithys albifrons* (Linné)
60	16	Army Ants; *Eciton burchelli* Westwood
97	10	Great Green Kingfisher; *Chloroceryle amazona* (Lath.)
	11	Tiny Emerald Kingfisher; *Chloroceryle americana* (Gmel.)
103	25	Gecko; *Thecadactylus rapicaudus* (Houtt.)
109	8	Howling Monkeys; *Alouatta seniculus macconnelli* Elliot

Page	*Line*	
113	7	Bower Bird; *Ptilonorhynchus violaceus* (Vieill.)
116	24	Cassava; *Janipha manihot* Kth.
126	20	Frog, Gawain; *Phyllomedusa* sp.
132	17	Marine Toad; *Bufo marinus* (Linné)
133	8	Scarlet-thighed Leaf-walker; *Phyllobates inguinalis.*
149	2	Attas, Leaf-cutting Ants; *Atta cephalotes* (Fab.)
151	12	Fruit Bats; *Vampyrus spectrum* (Linné)
152	11	King Vulture; *Gypagus papa* (Linné)
	11	Harpy Eagle; *Harpia harpyja* (Linné)
163	3	Ani; *Crotophaga ani* Linné
	7	Marine Toad; *Bufo marinus* (Linné)
164	19	White-faced Opossum; *Metachirus o. opossum* (Linné)
173	1	Attas, Leaf-cutting Ants; *Atta cephalotes* (Fab.)
	5	Hummingbird; *Phoethornis r. ruber* (Linné)
174	7	Tamandua; *Tamandua t. tetradactyla* (Linné)
175	1	Trogon; *Trogon s. strigilatus* (Linné)
	9	Tarantula Hawks; *Pepsis* sp.
181	17	Cicada larvæ; *Quesada gigas* Oliv.
182	5	Roaches; *Attaphila* sp.
231	26	Manatee; *Trichechus manatus* Linné
232	24	Crocodile; *Caiman sclerops* (Schneid.)
233	6	Jacana; *Jacana j. jacana* (Linné)
	8	Gallinule; *Ionornis martinicus* (Linné)
	9	Green Herons; *Butorides striata* Linné
	10	Egrets; *Leucophoyx t. thula* (Molina)
233	17	Kiskadees; *Pitangus sulphuratus* (Linné)
	19	Black Witch; *Crotophaga ani* (Linné)
	19	House Wren; *Troglodytes musculus clarus* Berl. and Hart
	22	Manatee; *Trichechus manatus* (Linné)
242	1	Jacana; *Jacana j. jacana* (Linné)
	3	Gallinule; *Ionornis martinicus* (Linné)
243	15	Mongoose; *Mungos mungo* (Gmel.)
246	11	Little Egret; *Leucophoyx t. thula* (Molina)
	14	Tri-colored Heron; *Hydranassa tricolor* (P. L. S. Mull.)
	15	Little Blue Heron; *Florida c. caerulea* (Linné)
249	14	White Egret; *Casmerodius egretta* (Gmel.)
250	10	Night Heron; *Nyctanassa violacea cayennensis* (Linné)
254	1	Giant Catfish, Boom-boom; *Doras granulosus* Valen.
256	6	Long-armed Beetle; *Acrocinus longimanus* (Linné)
276	10	Rufus Hummingbird; *Phoethornis r. ruber* (Linné)

Page	*Line*	
278	16	Tapping Wasp; *Synoeca irina* Spinola
280	10	Mazaruni Daisy; *Sipanea pratensis* Aubl.
	21	Trogons; *Trogonurus c. curucui* (Linné)
282	10	Quadrille Bird; *Leucolepis musica musica* (Bodd.)
284	3	Bubble Bugs; *Cercopis ruber*
289	16	Army Ants; *Eciton burchelli* Westwood

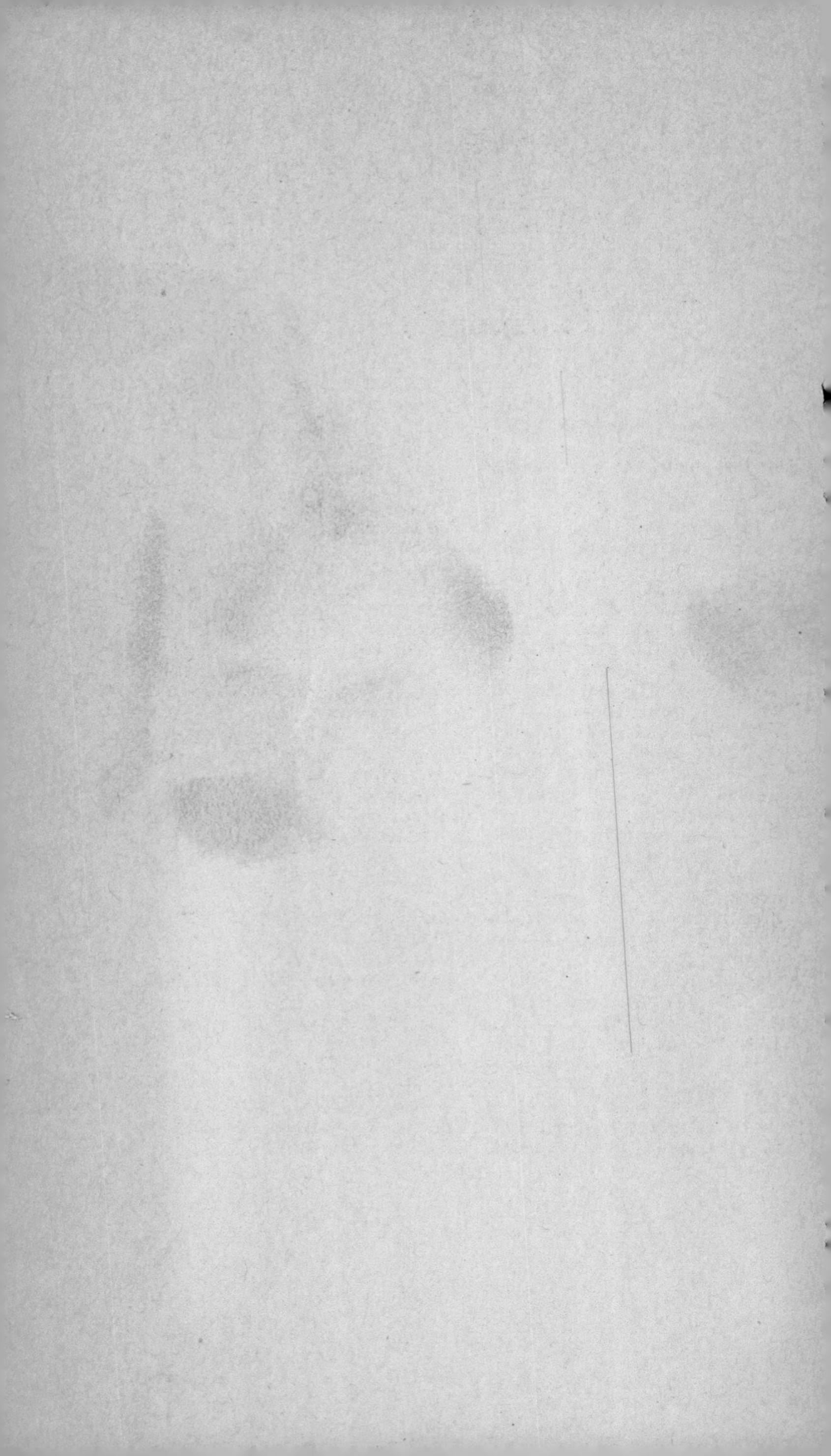

INDEX

G

H

I

J

K

L

M

N